5/-

RADIO RECEIVER
CIRCUITS HANDBOOK

ALSO BY E. M. SQUIRE

INTRODUCING
RADIO RECEIVER SERVICING

This book provides a concise introductory guide to the practical operation of a radio receiver, so that radio mechanics, service engineers, testers, and dealers may be able to obtain a sound working knowledge of receivers and servicing equipment in the briefest time.

Crown 8vo   6s. net   Cloth

"*Can be recommended as an excellent introduction to more advanced manuals.*"—WIRELESS AND ELECTRICAL TRADER.

"*Confidently recommended.*"—TECHNICAL JOURNAL.

"*There can be no doubt as to its conciseness, and a great amount of information has been packed into a small space. . . . Can be safely recommended.*"—WIRELESS WORLD.

PUBLISHED BY PITMAN

RADIO RECEIVER CIRCUITS HANDBOOK

CONTAINING PRACTICAL NOTES ON THE
OPERATION OF BASIC MODERN
CIRCUITS

BY

E. M. SQUIRE

SECOND EDITION

LONDON
SIR ISAAC PITMAN & SONS, LTD.
1943

Reprinted January, 1943

SIR ISAAC PITMAN & SONS, Ltd.
PITMAN HOUSE, PARKER STREET, KINGSWAY, LONDON, W.C.2
THE PITMAN PRESS, BATH
PITMAN HOUSE, LITTLE COLLINS STREET, MELBOURNE
UNITEERS BUILDING, RIVER VALLEY ROAD, SINGAPORE
27 BECKETTS BUILDINGS, PRESIDENT STREET, JOHANNESBURG

ASSOCIATED COMPANIES
PITMAN PUBLISHING CORPORATION
2 WEST 45TH STREET, NEW YORK
205 WEST MONROE STREET, CHICAGO

SIR ISAAC PITMAN & SONS (CANADA), Ltd.
(INCORPORATING THE COMMERCIAL TEXT BOOK COMPANY)
PITMAN HOUSE, 381-383 CHURCH STREET, TORONTO

MADE IN GREAT BRITAIN AT THE PITMAN PRESS, BATH
D3—(T.100)

PREFACE
TO THE SECOND EDITION

As the basic circuits used in modern radio receivers have not changed since this book was first written, extensive revision has not been necessary. Some of the circuits described have, nevertheless, been altered so as to make the book more representative of modern practice, and the text has been clarified in many parts.

A chapter has been added showing a selection of typical broadcast receiver diagrams ranging from a three-valve battery set to the "eight" valve receiver of the manufacturers (six valves plus mains rectifier and tuning indicator). The diagrams have been chosen with the object of including as many features as possible that have not previously been described but which are commonly used in modern receivers, such as negative feedback, variable selectivity, cathode ray tuning indicator, electron coupled triode pentode, etc.

E. M. S.

August, 1941.

PREFACE

THIS book has been written for the practical radio man. Its object is to present a summary of the general theory of the most important and commonly used circuits in modern radio receivers, without becoming in any way very technical. The endeavour of the writer has been to keep to "practical theory" only.

Notes are given on the best methods of operating the circuits described. These practical notes are written entirely from the point of view of modern practice, and they follow logically from the theoretical notes. Notes are also given on the faults in receivers that are likely to develop if the wrong operating conditions, or components, are employed when using the different circuits, and how to avoid these.

For the purpose of this book, the radio receiver has been split up into stages and sub-stages, as it is considered that in this way the outlines will be the most useful and concise.

It is hoped that the book will be found helpful to those many members of the radio industry and radio amateurs who "do not like too much theory" but who, nevertheless, are eager to know something of "how it works" and, perhaps more important still, the best way to make it work.

E. M. S.

LONDON,
November, 1937.

SHORT-WAVE RADIO
By J. H. REYNER.

An invaluable companion volume to *Modern Radio Communication*, and is recommended to all students of radio engineering as a reliable textbook on modern developments in the use of the short, ultra-short, and micro-waves.

Illustrated   10s. 6d. net   177 pp.

THE SUPERHETERODYNE RECEIVER
Its Development, Theory and Modern Practice.
By ALFRED T. WITTS, A.M.I.E.E.

Describes the operation of Superheterodyne Receivers in the clearest possible way. It makes every detail of the subject understandable and provides the essential working knowledge required by every keen amateur constructor, student of radio, and service engineer.

Cloth   5s. net   200 pp.

RADIO UPKEEP AND REPAIRS FOR AMATEURS
By ALFRED T. WITTS, A.M.I.E.E.

Enables the average radio receiver owner to diagnose the ordinary troubles of his wireless set and to remedy them himself.

Cloth   6s. 6d. net   215 pp.

RADIO RECEIVER SERVICING AND MAINTENANCE
By E. J. G. LEWIS.

A practical manual specially written to give the radio dealer, salesman, and the service man up-to-date and reliable assistance in the technical details of their work. A handy fault-finding summary, combined with the index, is a feature of the book.

Cloth gilt   8s. 6d. net   253 pp.

PROBLEMS IN RADIO ENGINEERING
By E. T. A. RAPSON, A.C.G.I., D.I.C., A.M.I.E.E.

A classified collection of examination questions set from time to time by some of the more important examining bodies in Radio Communication, together with some useful notes and formulae bearing on the different groups of questions and answers to those questions which are capable of a numerical solution.

In crown 8vo   4s. 6d. net   123 pp.

PITMAN

CONTENTS

CHAP.		PAGE
	PREFACE	v
I.	HIGH-FREQUENCY AMPLIFIERS	1
II.	DETECTORS	11
III.	LOW-FREQUENCY AMPLIFIER CIRCUITS	22
IV.	MULTIPLE VALVE CIRCUITS	32
V.	FREQUENCY CHANGER CIRCUITS	43
VI.	THE OUTPUT STAGE	54
VII.	PUSH-PULL AMPLIFIERS	62
VIII.	POWER SUPPLY CIRCUITS	75
IX.	COMPLETE MODERN RADIO RECEIVER CIRCUITS	87
	INDEX	103

A FIRST COURSE IN WIRELESS

Being a reprint in book form of a series of articles which appeared in *World Radio* under the title "The Radio Circle: for Beginners Only."

By "DECIBEL." In crown 8vo, cloth, 221 pp. **5s.** net.

WIRELESS TERMS EXPLAINED

Based upon a series of articles which were published in *World Radio* under the title of "A Wireless Alphabet."

By "DECIBEL." In crown 8vo, cloth, 80 pp. **2s. 6d.** net.

THERMIONIC VALVE CIRCUITS

By EMRYS WILLIAMS.
In demy 8vo, cloth, 174 pp. Illustrated. **12s. 6d.** net.

THE RADIO AND TELECOMMUNICATIONS ENGINEER'S DESIGN MANUAL

By R. E. BLAKEY, D.Sc.
In demy 8vo, cloth, 142 pp., illustrated. **15s.** net.

ELECTRIC CIRCUITS AND WAVE FILTERS

By A. T. STARR, M.A., Ph.D., A.M.I.E.E., A.M.I.R.E.
In demy 8vo, cloth gilt, 476 pp. **25s.** net.

CATHODE RAY TUBES

By MANFRED VON ARDENNE. Translated from the German by G. S. MCGREGOR and A. C. WALKER.
In demy 8vo, cloth gilt, 530 pp. **42s.** net.

A DICTIONARY OF ELECTRICAL TERMS
Including Electrical Communication.
By S. R. ROGET, M.A., A.M.Inst.C.E.
In crown 8vo, cloth gilt, 432 pp. **12s. 6d.** net.

Sir Isaac Pitman and Sons, Ltd., Parker St., Kingsway, W.C.2

RADIO RECEIVER CIRCUITS HANDBOOK

CHAPTER I

HIGH-FREQUENCY AMPLIFIERS

An amplifier valve is one that provides in the load device in its output circuit a voltage similar in waveform to that in its input circuit, but which is greater in value. The process of amplification takes place by virtue of the change in anode current effected by a change in control grid voltage. This varying anode current produces a voltage across the output or load impedance. The output voltage may be several hundred times greater than that applied to the input circuit.

A simple arrangement of amplifier valve is shown in Fig. 1. The input circuit is shown at "in" and the output circuit at "out," and the valve itself is, of course, a high-frequency pentode. The voltage ratio, termed the amplification, is that voltage developed across "out," compared with that across "in." The amplification is determined by a large number of factors and is generally approximately equal to the mutual conductance of the valve in amperes per volt multiplied by the dynamic resistance of the output circuit. From this it follows that the valve with the greatest mutual conductance will give the greatest amplification and, generally speaking, this is borne out by practice. It would also appear from what is stated above, that the most efficient circuits possible should be used in the output circuit, for the higher the efficiency of the output circuit, the greater will be the dynamic resistance of the circuit. The limiting factor in the use of pentodes for high-frequency amplification is, in fact, this question of the dynamic resistance of the output circuit, and in most

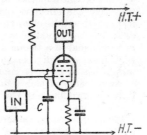

Fig. 1. Simple Amplifier Circuit

receivers the amplification of the stage is dependent upon the "goodness" of the output circuit.

There is, however, another factor that must be considered in connection with high-frequency amplifiers. This is the stability of the stage. It might be thought that as a valve with a screen grid is especially designed to have a very small inter-electrode capacitance, instability is not an important feature when these types of valves are employed for high-frequency amplification. The instability referred to here is that which results from the feed back of voltage from the anode circuit to the control grid circuit via the capacitance existing between the anode and the control-grid electrodes.

Now although the screen-grid valves and h.f. pentodes have a grid-anode capacitance that is only about one-thousandth that of a triode h.f. amplifier valve, it is found in practice that feed back does occur unless special precautions are taken. Feed back and consequent instability, i.e. the tendency to break into oscillation, are particularly noticeable when the valve is worked in such a way that it provides a high gain. The operating conditions for high gain are high anode and screen-grid voltage and low control grid bias. Under these conditions the mutual conductance is greatest and so is the effective amplification. The feed back is proportional to the effective amplification and the grid-anode capacitance. It will be clear that, although the grid-anode capacitance of a screen-grid valve or h.f. pentode is only a very small fraction of that usual in a triode h.f. valve, the actual feed back is not by any means negligible in comparison, owing to the effective amplification being so many times more.

As a result of this tendency to instability, there is a definite limit of amplification that can be obtained from straightforward h.f. amplifiers. It is usually a fairly simple matter to construct a satisfactory single stage of h.f. amplification, but not nearly so simple to construct a stable two-stage h.f. amplifier to give anything like the maximum possible overall amplification. In order to operate a two-stage h.f. amplifier of normal construction, it is generally necessary to take measures that have the ultimate result of reducing the actual amplification provided. In this respect the superheterodyne scores very considerably over the straightforward receiver. Referring again to Fig. 1, it will be apparent that quite apart from the capacitances existing in the valve, that tend to feed back a voltage from the output circuit to the input circuit, there still may be serious feed back if any stray capacitances are allowed to exist between these circuits. No matter how perfectly screened the electrodes are inside the valve, if the external circuits are poorly designed, the stage may be very unstable. Consequently, special steps have to be taken to ensure

HIGH-FREQUENCY AMPLIFIERS 3

effective screening of the components and circuits and to arrange that the wiring is such that there are no long parallel lengths of anode and grid leads.

The condenser shown at C is of vital importance in the operation of the amplifier, for by it the screen grid is tied to earth so far as any high-frequency currents are concerned. Although a fairly high voltage from the h.t. source is supplied to the screen grid, this voltage is a direct current voltage. For effective operation as an electrostatic screen, the screen grid must never be allowed

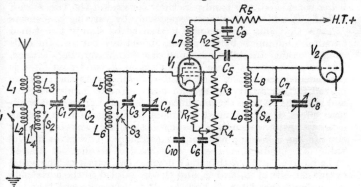

Fig. 2. H.F. Amplifier with Band-pass Input and Choke-coupled Output Circuit

to acquire an alternating or high-frequency voltage, and so long as the condenser C is connected, any high-frequency currents that reach the screen grid are passed directly to earth if the condenser has a low impedance for them. Condenser C must, therefore, have a capacitance of 0·1 to 0·25 μF. and be of the non-inductive type. It should be connected close to the screen grid.

These preliminary remarks are given to refresh the reader's memory on the main points in the operation of a high-frequency amplifier and also to save repetition in the descriptions of the practical circuits given below. These circuits, it should be mentioned, are not given in order of merit, as each type of circuit has its particular advantages.

Impedance-coupled Amplifier. An impedance-coupled amplifier is one in which the anode circuit consists of an untuned impedance. In high-frequency amplifiers this impedance is usually a high-frequency choke. In Fig. 2 this choke is seen at L_7 connected to the anode of the pentode valve.

One of the most important parts of high-frequency amplifier

circuits is the input circuit tuning arrangements. In order that the high-frequency signal may be selected without serious diminution of the higher sideband frequencies, it is desirable to use a band-pass filter. If ordinary coupled circuits without this so-called band-pass filter arrangement are used, there results in the reproduction a lack of the high notes, and music becomes very low pitched. The band-pass filter allows to pass a wide band of the frequencies in the signal at adequate amplitude and yet at the same time selects a station clear of interference. In the circuit being considered, the band-pass filter consists of the two sets of coils L_3 L_4 and L_5 L_6, tuned respectively by variable condensers C_2 and C_4. This arrangement is seen to be simply two tuned circuits. Coupling is by the mutual inductance between the coils usually arranged inside a screening case within coupling distance. The effectiveness of these coils to act as a band-pass filter is dependent largely upon their relative distance. As the coils are moved further apart, the system becomes more highly selective and at the same time the amplification falls off. For a good band-pass effect, the coils must be made to approach to a certain nearness until the combined characteristic of the two sets of tuned circuits has a flat top and fairly steep sides. This gives a level value of reproduction over a wide band of frequencies desired to be received and a satisfactory degree of selectivity. The aerial circuit consists of L_1 and L_2 and this is coupled to the aerial side of the band-pass filter L_3 and L_4. It will be apparent that the switching arrangement shown across the lower of each pair of the coils in this circuit is to short-circuit this part of the tuning system when the medium waves are being received and to open the short circuit for long waves and thus render operative the entire tuning system. Condensers C_1 and C_3 are the usual trimming condensers across the variable condensers. These level up the total of the stray capacitances in the circuit and components, so that correct ganging of the two tuned circuits forming the band-pass filter may be obtained.

The signal voltages applied to the control grid of the amplifier valve appear at the anode and owing to the impedance presented by the high-frequency choke L_7 an amplified voltage drop is produced across this component. This amplified voltage is passed by coupling condenser C_5 to the grid circuit of the following amplifier. When impedance-coupled amplifiers are used, it is preferable to use a tuned grid with the following amplifier for by doing so, increased selectivity and gain is obtained than if this circuit is made aperiodic by connecting a grid leak across the grid cathode of the second valve and doing without the tuned circuit. The tuned grid circuit is shown at L_8 L_9 and variable condenser at C_8.

HIGH-FREQUENCY AMPLIFIERS

The screen-grid voltage is provided by a potentiometer consisting of R_2, R_3, and R_4. In fitting this potentiometer, the value of R_2 should not be too great, otherwise the voltage at the screen grid will be lower than that required for satisfactory amplification. On the other hand, if R_2 is made too low, the voltage of the screen grid may be excessively high, and instability may result. The voltage drop along R_4, also in the screen-grid potentiometer circuit, is applied to the cathode for varying the amplification of the stage. By means of the tapping along R_4, the cathode may be given a varying positive voltage and, since the control grid is returned to earth, the grid is made correspondingly negative with respect to cathode. The resistance R_1 is to give a minimum value of grid bias so that the valve shall not be under-biased. At the same time, R_1 is necessary to prevent instability when the minimum bias is applied by the volume control potentiometer. If instability is experienced in the working of an h.f. amplifier stage, it will frequently be found that increasing the value of the minimum bias resistance R_1 will overcome this difficulty. This measure will at the same time reduce the maximum sensitivity.

R_5 is partly for decoupling and partly to drop the full h.t. voltage down to that required by the anode. Its value is dependent entirely upon the value of the h.t. supply and may frequently be done without. The decoupling condenser C_9 should have the usual value of 0·1 μF.

The arrangement shown in Fig. 2 is a very simple one so far as the coupling between the amplifier valve and the following stage is concerned. If the high-frequency choke L_7 is of reputable make, the circuit will be quite stable in operation and satisfactory results will be obtained. A common defect in ordinary h.f. chokes is that the internal capacitance is large. This by itself is not a serious defect, for it means that there is an extra capacitance across the variable condenser C_8. The effect of this can be easily overcome by reducing the amount of the trimmer condenser C_7 in circuit.

If the frequency at which the choke's stray capacity resonates with the choke's inductance comes within the band of frequencies tuned by the amplifier, the choke becomes in effect a tuned anode circuit. The amplification then becomes peaky and quite likely the whole stage will go into self-oscillation. To avoid this defect, the choke should be so designed that its natural frequency is lower than the lowest carrier frequency the amplifier is to tune.

The value of the coupling condenser C_5 is of some importance. Its value should not be greater than is necessary to operate the circuit satisfactorily, and normally a value of 0·0001 μF. will meet the purpose. Frequently, a value of 0·0005 μF. may be used, but in other cases a value of 0·0003 μF. is found to be the most satisfactory.

RADIO RECEIVER CIRCUITS HANDBOOK

Transformer-coupled Amplifier. A transformer-coupled amplifier is one in which the coupling between the output of the amplifier and the input of the subsequent stage is effected by means of a transformer. This coupling transformer may have a tuned primary, a tuned secondary, or may be untuned. Generally the secondary is tuned and the primary is untuned when the amplifier is used for high-frequency voltages. When the stage is part of a superheterodyne intermediate-frequency amplifier, it is

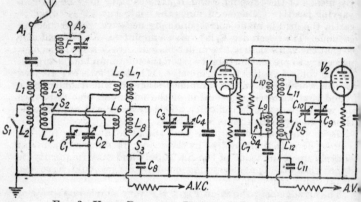

Fig. 3. Using Band-pass Input and Transformer-coupled Output Circuit

the common practice to tune both primary and secondary windings of the transformer and to couple them in such a way that a band-pass filter effect is produced. The coupling transformer referred to here is that shown in Fig. 3 at L_{10} L_{11} and L_9 L_{12} respectively. In the circuit shown in Fig. 3 there are two alternative connections for the aerial. That shown at A 1 is the normal connection of the receiver and that at A 2 is the connection used when interference from a strong station is being experienced.

The filter circuit F is a rejector of voltages at the frequency to which it is tuned. At the particular frequency to which the inductance and the capacity forming the rejector resonate, a very high impedance is produced and this results in voltages at that particular frequency having their path through the filter blocked. It is convenient to have a rejector circuit in the position shown at F, for in many districts serious interference is produced by one particular transmitter. All that is necessary is to have a suitable coil and a preset condenser and to tune the rejector circuit by the condenser to that point at which the interfering signal is a

HIGH-FREQUENCY AMPLIFIERS

minimum. The rejector circuit can then be left and the switch inserted in the aerial lead so that in the position 1 the straightforward connection is made between aerial and receiver and in position 2 the rejector is thrown into circuit during periods of strong interference by the particular transmitter to which F is tuned. The filter F can, of course, be made to tune to any station as desired.

Coupled to the aerial circuit is a band-pass filter arrangement, but in this circuit the coupling between the elements forming the filter is somewhat different from that employed in the filter of Fig. 2. In the arrangement shown in Fig. 2 the tuning coils themselves are directly coupled. In the arrangement shown in Fig. 3, however, the main tuning coils themselves are not directly coupled, but instead a proportion of the requisite inductance is made as a separate coil and used for providing the necessary degree of coupling to produce the band-pass effect. The main tuning coils in the primary circuit, for example, are L_3 and L_4 and the coupling coils are L_5 and L_6 respectively. The current set up in the coils L_3 and L_4 by the aerial coils passes round the coupling coils L_5 and L_6, and induces a voltage into the secondary coils L_7 L_8. The flow of current round the circuit L_7 L_8 C_8 and C_3 produces a voltage on its own account, and this is back-coupled via the coupling coils L_5 and L_6 to the main coils L_3 and L_4, which are tuned by the variable condenser C_2. These two circuits thus act in combination and are designed so that a fairly flat-topped band-pass is provided. This arrangement of band-pass filter is known as an inductively-coupled filter, the coupling coils L_5 and L_6 being the sole inductive links. It should be mentioned that there are other forms of inductively coupled band-pass filter. For example, instead of using L_5 and L_6 for coupling, the earth return of the main tuning inductances L_3 L_4 and L_7 L_8 could pass through an inductance coil which is common to both the primary and secondary circuits of the filter. This would give the desired effect and is, in fact, sometimes used as a band-pass filter coupling.

The amplified version of the input voltages appears across the primary coils L_9 and L_{10} forming the primary of the output transformer. The flow of h.f. anode current in these coils induces voltages in the secondaries L_{11} and L_{12} of this transformer, and these secondaries are tuned by variable condenser C_9 to resonance. The output transformer in this case need not have a step-up ratio, but can be of the 1 : 1 type, i.e. it may have the same number of turns in the secondaries L_{11} L_{12} as are in the corresponding primaries L_{10} and L_9. A voltage step-up will be provided, however, if there is a step-up ratio in the same way as in any other transformer. Difficulties arise, however, if it is desired to use a high step-up ratio, and in practice a transformer ratio of 4 : 1 is about the highest one meets. Although the secondary is tuned and the primary is untuned, this does not complicate the operation of the transformer for, owing to the coupling between the windings, the act of tuning the transformer secondary tunes also the primary. This gives to the primary circuit a satisfactory dynamic resistance if the transformer ratio is not too high. For practical purposes, therefore, an untuned primary and a tuned secondary of the h.f. transformer can be considered satisfactory for ordinary selective reception. If, however, the band-pass effect obtained by the input to the amplifier is desired to be accentuated, then the primary may be tuned also and the coupling between the two windings adjusted for the best position to give a band-pass filter effect. This latter arrangement is the one usually encountered in the intermediate-frequency amplifier of a superheterodyne receiver in which only one frequency has to be tuned.

The two valves shown in Fig. 3 should be of the variable mu type, and they may be either screen-grid valves or pentodes. It is seen that a.v.c. is used, C_8 and C_{11} being the decoupling condensers usual for filtering out the modulation component in the control voltage. In this circuit the screen grid of the amplifier valve is shown to be supplied with h.t. through one resistance only from the main h.t. supply instead of the potentiometer illustrated in Fig. 2. Whether a potentiometer is used or not will depend partly on the type of valve, but primarily on the desire of the user. It is becoming the more common practice, however, to use the single resistance as shown in Fig. 3. It might be mentioned that should any tendency to instability become apparent, this may be sometimes reduced by connecting a resistance of about 1 000 ohms next to the screen grid.

Tuned Anode-coupled Amplifier. This type of circuit is one in which there is a tuned circuit connected to the anode of the valve, voltages from this tuned circuit being applied to the following stage through a coupling condenser. This arrangement is distinct from that of a doubly tuned transformer-coupled amplifier for in this latter case the voltage is applied to the following stage by means of the secondary transformer winding. The difference between the two circuits will be apparent from Fig. 4. The tuned anode circuit is L_7 L_8 with variable condenser C_6. This tuned anode circuit may be considered as taking the place of the choke L_7 shown in Fig. 2 and, in fact, the operation of this type of amplifier is similar in many respects to that of the output circuit of Fig. 2. The important difference, however, between the two output circuits is that the arrangement of Fig. 4 is generally more selective and if carefully designed will provide a higher amplification owing to the greater dynamic resistance of the tuned anode

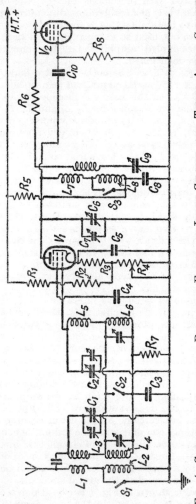

Fig. 4. Capacitively-coupled Band-pass Filter as Input Circuit, and Tuned Anode Circuit

circuit as compared with the choke coil. On the other hand, a tuned anode circuit is considerably more difficult to operate owing to the tendency to instability that is found to be present. Tuned anode circuits are not used so extensively in modern receivers as the transformer-coupled amplifier, but the circuit has many merits as already mentioned.

Referring to Fig. 4 it will be noticed that a band-pass filter is again used between the aerial circuit and grid cathode path of the amplifier valve V_1. In this case the filter is of the capacitively coupled type. The two sets of coils L_3 L_4 and L_5 L_6 are screened from each other and have no inductive connection whatever. In the common earth return of both sets of coils, however, there is a condenser C_3 which acts as the coupling condenser. When voltage is induced into coils L_3 L_4 by the aerial coils L_1 L_2, a current is set up which passes round the coils, variable condenser C_1, coupling condenser C_3 to earth. The passage of the current round this circuit sets up a voltage across the coupling condenser C_3, of course, and since this condenser is common to the secondary circuit consisting of L_5 L_6, variable condenser C_2, and coupling condenser C_3, a corresponding voltage is set up in this circuit also. Thus, although the coils are totally separate, the two circuits have a common element C_3, and owing to this the resonant characteristics of the two circuits are combined to form a resultant band-pass characteristic of the type already outlined in connection with the previous circuits.

Capacitance-coupled band-pass filters are used extensively in straight receivers, but they suffer from the drawback that the characteristic varies considerably at different points of the tuning range. The impedance of the coupling condenser C_3 varies inversely as the frequency of the current passing through it. Consequently, at the lower frequency ends of the tuning ranges the voltage produced across the coupling condenser will be many times greater than that produced across it at the higher frequency ends. This causes a considerable alteration in the amplification of the stage and adds to the disadvantages of this type of filter. This disadvantage can be overcome by the use of a coupling that combines the capacitance coupling with the inductive coupling shown in the previous circuits. This type of circuit is known as a mixed coupled band-pass filter. C_3 has a capacity of about 0·025 μF. and is shunted by a resistance R_7 (a few thousand ohms) to level up the coupling and complete the grid d.c. circuit.

The second valve shown in Fig. 4 is a grid detector valve, the operation of which is described in the next chapter. The insertion of R_6 (about 300 ohms) enables smoother reaction to be obtained.

CHAPTER II
DETECTORS

The general conditions under which the detector circuit operates can be visualized from an examination of Fig. 5. The purpose of the detector is to cut off one-half of the received signal waveform and to render the resultant voltage reproducible by a low-frequency amplifier or loudspeaker. The signal voltage as received by a radio receiver is incapable of operating a loudspeaker owing to the fact that the alternations in voltage are so rapid that the loudspeaker cannot respond to them. The detector, by cutting off one-half of these signal voltages and at the same time converting the detected half-waves into series of groups, corresponding to the signal modulations at the transmitter, thus performs an essential feature in all radio receivers.

The operating curve of Fig. 5 applies to all grid and anode bend detectors. The particular curve shown is not the ideal one however. It is, nevertheless, a general one and the detectors described later in this chapter operate in a similar manner to that shown. The differences in operation of the various detectors outlined below are brought about mainly by the voltages applied to the electrodes and by the position and extent of the bends shown round about the points P and S.

A typical radio signal voltage is shown at VS in Fig. 5. This curve represents a carrier wave bearing a modulation corresponding to the signal transmitted. The detector is operated in such a way that the received signal voltage fluctuates on either side of the point P. The current in the output of the grid or anode detector follows the form shown on the right of the Fig. and the individual half-waves can be traced. The loading device connected in the output of the detector responds to the average value of the detected half-waves, and this average value corresponds to the actual signal originally impressed on the transmitted carrier wave.

The points to be noted in connection with the operating curve of the detector are—

(1) The position of the point P. This point should be on a bend of the operating curve so that as the signal voltage fluctuates on either side a change in average value of the output current is effected.

(2) The saturation point S limits the magnitude of input voltage that can be applied to the detector for at this point the

curve bends over and does not reproduce any increase in voltage input, i.e. distortion is set up.

(3) The distance of S from P. It is quite clear that the greater this distance is, the larger must be the signal voltage input before the upper bend is reached.

It can be seen that if a very small voltage input is applied to the detector at point P on its operating curve, the fluctuations in the output current will be almost exactly the same as the input

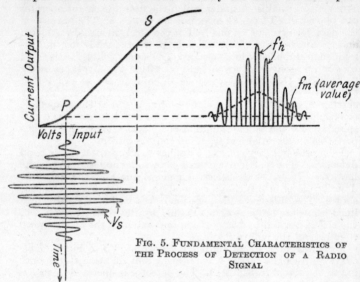

FIG. 5. FUNDAMENTAL CHARACTERISTICS OF THE PROCESS OF DETECTION OF A RADIO SIGNAL

voltage owing to the similarity in the slopes of the curve close to point P on either side. It follows, therefore, that weak signals will be detected very inefficiently unless the detecting device has a characteristic which is sharply curved at its operating point P. Conversely, if the detector has a very round curve near its operating point, then it will require a correspondingly larger voltage input before it will detect the signals satisfactorily.

It is seen from the detector curves to the right of Fig. 5 that there are two separate waves to be dealt with by the circuits following the detector. One of these waves has a frequency f_h. This is the frequency of the carrier wave, and once it has set up the group frequency, i.e. the desired modulation frequency, f_m, serves no further purpose except for reaction. The carrier frequency f_h is a serious obstacle to the operation of the subsequent

DETECTORS 13

stages in the receiver if it is allowed to penetrate into them. In all receiver circuits it is necessary, therefore, to provide means for ridding the receiver of f_h and of extracting the modulation component f_m as efficiently as possible. In practice, it is found that a small condenser connected across the output circuit of the detector is sufficient to by-pass the carrier frequency f_h without interfering with the modulation frequency f_m. On the next page is described how f_h is used for reaction.

The above are the general conditions for operating detectors, and a perusal of Fig. 5 will help to clarify the explanations that follow of the practical grid and anode bend detector circuits.

Grid Detector. The grid detector operates on the grid voltage-grid current curve. The shape of the curve is similar to that of the operating curve in Fig. 5, the grid current being of the order of micro-amperes. The valve is worked on the correct operating point P by proper dimensioning of the grid resistance and by the connection of the cathode end of that resistance to a source of positive voltage in the case of battery valves, or directly to the cathode in the case of mains valves.

When the valve is first connected into the circuit a certain value of current will flow from the cathode to the anode and in so doing set up a current between cathode and grid. This electron current then flows back to the cathode through the grid leak R, Fig. 6, and thereby sets up a voltage which at the grid is slightly negative with respect to the cathode. The negative voltage of the grid is dependent upon the magnitude of the current flow and the value of R. For a given valve, therefore, the grid leak R determines the operating point worked upon for any given value of anode voltage. The grid current cannot flow back to the cathode through the tuning inductance L_1 owing to the position of C_2, which blocks this path to the direct current from the grid.

With the valve, Fig. 6, properly adjusted to the operating point on the grid voltage-grid current curve, the operation of the circuit is as follows. The signal voltage is applied to L_1 which is tuned by C_1 to the carrier frequency of the signal. The voltage set up across the tuned circuit L_1 C_1 is applied via grid condenser C_2 to the grid of the detector valve. The grid voltage is, therefore, fluctuated on either side of the steady voltage it attained before the signal voltage was impressed on it. During the positive half-cycles of signal voltage the electrons (which are negative in sign) are attracted increasingly by the grid, and this accumulates in the condenser C_2 a negative charge. During the negative half-waves of the impressed signal voltage, the negative charge in C_2 passes through R back to cathode, and this added to the effect of the negative half-wave itself, enhances the detecting effect of the valve. It is thus seen that in addition to the operation of the valve on a suitable point of its operating curve, the effect of the grid condenser C_2 and leak R is to make the detector more efficient.

The voltage fluctuations on the grid are amplified at the anode owing to the amplification due to the valve itself, and in the anode circuit, consisting of the h.f. choke HFC, and the primary winding of the l.f. transformer T there will appear magnified voltages of the detected signal. The low-frequency currents flowing in the primary winding of T induce l.f. voltages in the

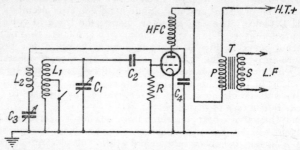

Fig. 6. Circuit of a Grid Detector

secondary winding S, and these are applied to the subsequent l.f. amplifier.

The high-frequency choke blocks the path through the primary of T for the high-frequency component at the anode and sends this component back to earth through two paths. One route to earth is through the inductance L_2 and reaction condenser C_3. The high-frequency component in passing through L_2 sets up a corresponding voltage in L_1 and this is again applied to the grid of the valve and appears amplified at the anode. The amount of reaction current, as this is called, that is allowed to pass through L_2 to earth is determined by the reaction condenser C_3, the larger the effective value of C_3 in circuit the easier it being for the reaction current to pass to earth. If too much reaction is employed, the circuit will go into oscillation and it will be impossible for the detected signal to be extracted from the output circuit.

The other route for the high-frequency component from the anode to cathode is through the by-pass condenser C_4. This condenser is necessary to ensure that when the reaction condenser C_3 is turned to a very low value, i.e. little or no reaction is being used, there is still an alternative path for the high-frequency component in the signal from the anode of the detector valve back to earth. If the high-frequency component forced its way through

DETECTORS

to the low-frequency circuits, satisfactory reception would be impossible.

In practice, the value of grid condenser C_2 employed varies from 0·00005 μF. to 0·00025 μF., and the value of grid-leak resistance R from 100 000 ohms to $2\frac{1}{2}$ megohms. The value of the components used governs very considerably the quality of the low-frequency signal reproduced by the receiver and also the magnitude of the voltage that may be detected without intolerable distortion. If it is desired, for example, to reproduce a strong signal at good quality, then the value of C_2 should be about 0·0001 μF. and R should be about $\frac{1}{4}$ megohm. Keeping C_2 at 0·0001 μF., it is found that as R is increased in value the reproduction of the higher musical frequencies becomes less satisfactory. On the other hand, the sensitivity of the detector is increased by using a higher value of R owing to the fact that there is less damping on the tuned circuit L_1 C_1. For the same reason the selectivity is enhanced by the use of a high value of grid leak R. It is thus clear that as the value of R is reduced, the range of musical frequencies passed by the detector is increased and correspondingly the selectivity and sensitivity of the detector become poorer.

For high sensitivity the value of C_2 may be 0·00025 μF., and the value of grid leak R may be 2 megohms. With the latter values, however, apart from the reduced quality of the reproduced signal, there occurs the unpleasant distortion that is brought about by the inability of the detector valve to reproduce at the correct strength the very loud passages as compared with the weaker ones.

The voltage applied to the anode of the grid detector valve also influences very considerably the efficiency of the detector valve. When the high values of grid condenser and grid leak are employed, a low anode voltage only is required. On the other hand, when small values of grid condenser and grid leak are used, it is imperative, if the utmost is to be made of this system, that as high an anode voltage as practicable, up to the maximum rated voltage of the valve, should be used. The value of the condenser C_4 varies in practice very considerably depending on the actual construction of the set and the type of valve used as a detector. In some receivers a value of 0·0001 μF. is found sufficient, whereas in others a value of 0·002 μF. is found to give by far the best results from the point of view of quality reception and sensitivity of the detector valve.

Anode Detectors. For anode bend or anode detection the valve is worked on the bottom bend of the grid voltage-anode current curve. The general shape of this curve is similar to the operating curve given in Fig. 5 for all types of multi-electrode valve, and it is a fairly simple matter to apply such a grid bias to the valve that the most satisfactory point on the lower bend of the curve is operated on. It might be mentioned that the curved part of the grid voltage-anode current characteristic is considerably more extensive than the grid voltage-grid current curve, and this results in the anode detector being much less sensitive than the grid detector. However, the anode detector has the advantage that it applies less damping to the tuned input circuit and thereby enables a receiver of higher selectivity to be designed. The anode

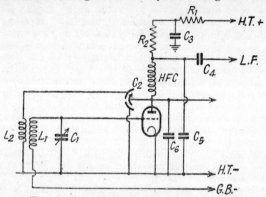

FIG. 7. ANODE BEND DETECTION CIRCUIT

detector will also handle a considerable input voltage and is satisfactory for fairly high gain receivers.

A circuit using the anode detector is given in Fig. 7. The operating point on the grid voltage-anode current curve is adjusted by means of a grid biasing battery GB. The most suitable bias voltage is easily to be found from an examination of the curves relating to the valve being used, the suitable bias voltage being that corresponding to the middle of the lower bend. This will give the greatest variation in average output current for a small fluctuation in input grid voltage. The process of detection when the correct bias has been applied to the anode detector is then very similar to that shown in Fig. 5.

The operation of the circuit shown in Fig. 7 is as follows. To tuned circuit L_1 C_1 are applied the signal voltages to be detected and, when this circuit is tuned to resonance by C_1, a voltage at carrier frequency is applied to the grid of the valve. The voltage at the grid is thereby varied on either side of the bias applied to it by GB and, in the manner described in connection with Fig. 5, a rectified voltage appears in the anode circuit of the detector valve. The high-frequency choke bars the way for the high-frequency component and this is passed back to earth partly through the valve by-pass condenser C_6 and partly through the differential reaction condenser C_2 and the reaction winding L_2. These components perform a similar function to L_2, C_3, and C_4 as outlined in connection with Fig. 6. The use of a differential condenser C_2, however, has the advantage that there is always a certain amount of capacitance across the valve and also that tuning is not upset to such an extent as when an ordinary two-electrode condenser is employed for providing the control of reaction.

The low-frequency component of the detected current passes through to the anode resistance R_2 and sets up a corresponding voltage therein. This is applied through low-frequency coupling condenser C_4 to the control grid of the subsequent stage. Resistance R_1 is, in conjunction with C_3, to decouple the anode circuit from the h.t. supply source by offering a low impedance path through C_3 for signal currents. Thus, the signal currents pass round C_3 to cathode in preference to the high resistance path through R_1 and the h.t. supply. This prevents undesired low frequency feedback commonly known as "motorboating," which sometimes occurs if decoupling is not used.

The values of the components used in anode detection are similar to the corresponding components used in grid detection as already outlined. The additional components used in the circuit now being described are the two resistances R_1 and R_2 and the condensers C_3 and C_4. The value of R_1 is usually from 10 000 to 20 000 ohms and C_3 from 1 to 4 μF. Satisfactory decoupling can frequently be obtained with 10 000 ohms and 4 μF. for R_1 and C_3 respectively, but it not infrequently happens that motor boating occurs in a receiver unless a capacity of 4 μF. is used for C_3 and up to 20 000 or even 30 000 ohms for R_1. The value of R_2 is dependent on the type of valve used as the anode detector and usually is equal to about three times the anode a.c. resistance of the detector valve. This particular arrangement is in effect a resistance capacitance-coupled amplifier, which is described in the chapter on low-frequency amplifier circuits. It is sufficient to mention here that C_4 is the coupling condenser for the low-frequency voltage and its value should not be too low, otherwise the higher modulation frequencies will be cut off. A normal value for C_4 is 0·02 μF., but in practice this value can be varied between 0·001 μF. and 0·1 μF.

Pentode as Anode Detector. When a more sensitive detector is required, the pentode is used and, owing to its much higher sensitivity, it provides a high degree of amplification. At the same time, mention should be made that the pentode is liable to produce a greater distortion than the triode valve as a detector, and this should be borne in mind when the pentode detector is being considered.

A satisfactory circuit is given in Fig. 8. In effect this circuit is almost exactly the same as that in Fig. 7 with the necessary alterations for rendering the pentode valve operative. The extra condensers shown at C_2 and C_3 are for trimming the particular inductances to the correct value for alignment purposes with the

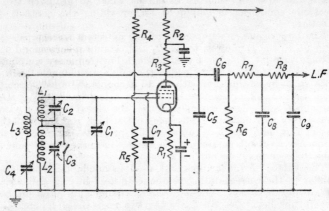

FIG. 8. PENTODE-ANODE DETECTION, SHOWING METHOD OF FILTERING OUT THE CARRIER FREQUENCY COMPONENT

other tuned circuits to which the detector input circuit is ganged. The grid bias for working the valve on its bottom bend is obtained by means of the biasing resistance R_1. This type of bias is explained in the chapter on power supply. The screen grid voltage is provided by the connection of the screen to the junction of R_4 and R_5, and the condenser C_7 is to maintain the screen grid at earth potential as regards any high-frequency current that may reach it.

The values of the resistances R_2 and R_3 are much larger than the anode resistances employed with the triode-anode detector of Fig. 7, owing to the much higher anode a.c. resistance of the pentode valve. A suitable value for R_2 is 30 000 ohms and for R_3 250 000 ohms. The use of these high anode resistances reduces very considerably the voltage applied to the anode, with the result that, when using a pentode as anode detector, a very low anode voltage is usual. Practical examples of operating voltages are—

DETECTORS 19

> Anode voltage . . . 45 volts.
> Anode current . . . 0·15 milliamps.
> Screen-grid voltage . . 25 volts.
> Screen-grid current . . 0·05 milliamps.

It is thus seen that the pentode anode bend detector is very economical from the point of view of current consumption.

The only other point that warrants comment in respect of Fig. 8 is the by-pass condenser across R_1. The value of this condenser may be from 1 to 10 μF. The actual value used will depend upon personal taste with regard to the quality of the reproduced music, the higher value giving a greater proportion of the lower musical frequencies. An electrolytic condenser can be satisfactorily employed in this position.

Diode Detectors. The diode or two-electrode valve is used very extensively in radio receivers as a detector, principally on account of the very efficient detection that is provided by it. The effect of the load resistance is to prevent the valve being operated on the part of the curve much to the right of P (Fig. 5). Consequently it is possible to receive and detect very high signal voltages without noticeable distortion. High-gain receivers such as multi-stage superheterodynes almost invariably employ diode detection owing to the large signal handling capabilities of the diode valve. Normally it is not necessary to apply any external voltage to the diode in order to operate it at the most suitable point of its characteristic, i.e. at the point P of the curve shown in Fig. 5. It is usually satisfactory to connect the diode directly to the tuned input circuit and to take off from the diode load the low-frequency component, f_m.

In Fig. 9 is seen a simple circuit arrangement for using a single diode as detector. The usual tuned circuit $L\ C_1$ is brought to resonance by the variable condenser at the frequency of the incoming carrier wave and the voltage across the terminals is applied between anode and cathode of the diode. During the peaks of carrier frequency voltage the diode anode is at a positive potential and there is allowed to flow from cathode to anode an electron current emitted by the cathode. This current flows from the anode through the tuning circuit and diode load resistance R back to cathode. The current thus passing round the circuit is that corresponding to f_h of Fig. 5. The high-frequency component is by-passed from the diode load R by the condenser C_2, the latter presenting a much lower impedance to the high-frequency component than R. The low-frequency component along the diode load R is passed to the subsequent low-frequency stage by the coupling condenser C_3. The coupling condenser, of course, prevents the direct voltage component along R from getting into the l.f. circuit.

The flow of electron current in the external circuit from the diode anode to cathode produces a voltage drop along R that makes the anode negative with respect to the cathode. The negative voltage of the anode is dependent upon the current flow, so the stronger the input signal, the more negative becomes the anode. If the value $C_2 R$ (the time constant of the load circuit) is suitably chosen, the anode acquires a mean negative voltage dependent upon the peak amplitude of the incoming carrier f_h. This negative voltage is maintained by the current flow at the peaks of the carrier frequency voltage. The modulation component f_m fluctuates about this average negative voltage, and has to be extracted for purposes of l.f. amplification and sound reproduction. For a.v.c., the average negative anode voltage (d.c.) due to the carrier frequency voltage is employed.

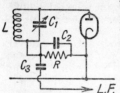

FIG. 9. SIMPLE DIODE DETECTOR CIRCUIT

The circuit shown in Fig. 9 is quite satisfactory in practice. Half-wave detection is obtained by the use of this circuit. This type of detection is similar to that already outlined in connection with the other circuits referred to in this chapter. The disadvantage of diode detection as compared with grid or anode detection is that there is no amplification from the diode valve, whereas when a triode or a pentode valve is used a considerable amplification is simultaneously provided by the detector valve. On the other hand, with a high-gain receiver, the amplification provided by the detector is not always necessary and in this case the diode detector is of great value by providing a higher quality of reproduction.

So long as the value of R is several times greater than the anode a.c. resistance of the diode itself, the damping of the diode and load circuit on LC_1 is approximately equal to $R/2$. A value from $\frac{1}{2}$ to 1 megohm for R is usually satisfactory, and 0.0001 μF. for C_2 is suitable. If C_2 is increased much above this figure, reproduction of the higher notes falls off. In any shunt circuit to $C_2 R$, such as is used for a.v.c. filtering (see p. 32), the shunt resistance should have a value of not less than 1 megohm.

Push-pull Diode Detection. In the outline given above of diode detection it was mentioned that half-wave detection was effected by the circuit shown in Fig. 9. By using two diodes it is practicable to detect both halves of the carrier wave and thus obtain full-wave detection. This is an advantage, as a higher efficiency of detection is thereby obtained, although the disadvantage that no gain is provided by the detector stage is still present.

A satisfactory circuit for push-pull diode detection is given in

Fig. 10. As the diode-anodes D_1 and D_2 become alternately positive with respect to the cathode owing to the impression on those anodes of the high-frequency voltage in the tuned circuit $L\ C_1$ provided by the incoming signal, these anodes pass current. The anodes D_1 and D_2, it is noticed, are connected to opposite ends of the tuned input circuit. The result of the alternate passing of current is therefore that the electron current passes through the two halves of the tuning inductance in opposite directions. The diode load R is connected to the electrical centre point of L and the two currents from the respective diode-anodes pass down the load resistance in the same direction. The direction of the current flow is indicated by arrows in Fig. 10. In practice it is found rather difficult to work this circuit entirely satisfactorily owing to the difficulty of tapping the true electrical centre of L. The circuit, nevertheless, may be used with a special design of input transformer designed to overcome this difficulty. As the high-frequency component flowing along R is now double the frequency of that usual in a half-wave detector, it becomes increasingly easy to by-pass the high-frequency component by means of C_2.

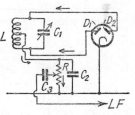

Fig. 10. Push-pull Diode Detector

The values of the components for operating the circuit shown in Fig. 10 are similar to those required for the circuit of Fig. 9. The value of C_2 in Fig. 10 need not, of course, be so large as the corresponding condenser in Fig. 9 owing to the fact already mentioned that the frequency of the high-frequency component is double that corresponding to the circuit of Fig. 9.

It is practicable to use a double diode valve for half-wave detection. In this case both diodes are connected together and the circuit employed is otherwise the same as that shown in Fig. 9. Strapped diode detection, as this arrangement is called, is used extensively in broadcast reception. Other circuits for using the diode detector are given in the chapter on multi-valve circuits. One example of using strapped diodes as a detector are given in Fig. 17.

CHAPTER III

LOW-FREQUENCY AMPLIFIER CIRCUITS

THE requirements for satisfactory low-frequency amplification are similar in many respects to those applicable to the high-frequency amplifier as already outlined in Chapter I. That is to say the valves must be worked on the straight part of their characteristics, the slope of which, for the highest gain, should be as high as possible. In the low-frequency stages of a receiver, however, the signal voltage is very much greater owing to the effective amplification of the preceding stages in the receiver.

In most cases the high-frequency component in the signal will have been completely eliminated by the time the signal voltage

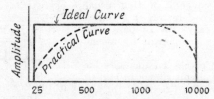

FIG. 11. CHARACTERISTIC CURVES OF AN L.F. AMPLIFIER

reaches the low-frequency amplifier. It does not follow, however, that no steps need be taken to eliminate stray carrier frequency voltage and some of the circuits given in this chapter include a condenser connected across the input circuit to the amplifier for this particular purpose. It should be remembered in considering the circuits given here that the frequencies of the voltages being considered are low, i.e. from 25 up to 5 000 or 10 000 cycles per second.

The ideal low-frequency amplifier is one that responds equally to frequencies throughout the whole of the audible range mentioned above. In Fig. 11 is shown a curve depicting the characteristic of such an amplifier. Along the base line are marked frequencies in cycles per second and the ordinate represents the amplitude of the output voltage. The curve is seen to rise rapidly at 25 cycles to the maximum output voltage and then to keep perfectly level until the highest desired audio-frequency has been reached, at which point the curve falls again abruptly to zero.

LOW-FREQUENCY AMPLIFIER CIRCUITS

The lowest frequency in that curve, namely, 25 cycles per second, is that at which the human ear becomes unresponsive to notes. Any frequencies lower than this, therefore, will be discerned as separate impulses. The upper limit is governed by the design of the receiver. Highly selective circuits usually do not pass frequencies higher than 5 000 cycles, and in many cases the upper frequencies are considerably lower than this. The background mush due to heterodyning and general interference in the receiver can usually be considerably diminished by the use of an upper cut-off at somewhere in the region of 5 000 cycles per second, and, in the popular type of broadcast receiver, reproduction is seldom adequate above this frequency.

A practical output of a low-frequency amplifier is shown in dotted lines and it is seen to fall rather rapidly as the 10 000 cycle point is reached. It should be mentioned here that the amplitude shortcomings of the output of an audio-frequency amplifier may not be due entirely to the low-frequency stages. In fact, as already mentioned in the chapter on high-frequency amplifiers, it frequently happens that the higher modulation frequencies in the broadcast signal are greatly reduced in amplitude by the high-frequency circuits when high selectivity is employed. At the same time, unless the low-frequency amplifier is very carefully designed, it too will diminish the amplitude of the modulation frequencies both near the upper and the lower limits as indicated by the broken line curve in Fig. 11.

It is practicable to design low-frequency stages that will pass up to 10 000 cycles and even higher audio-frequencies without serious attenuation. For selective broadcast reception of foreign stations on the medium and long waves such an amplifier is unnecessarily good since the highest frequencies transmitted by stations working on those waves are often not higher than about 4 500 cycles owing to the limitations imposed by the frequency spectrum allocated to each station by international agreement. On the ultra-short waveband, however, there is no such limitation, and it is found practicable both to transmit and receive modulation frequencies up to something approaching 20 000 cycles per second. For quality reception on the longer waves it is preferable to confine reception to a small number of good quality transmitters for which a high degree of selectivity is not needed.

Resistance Capacitance-coupled Amplifier. Resistance capacitance-coupled amplifiers are used very extensively in broadcast receivers owing to the high quality of reproduction that is obtainable by their use and also to the ease with which such an amplifier can be designed. The highest modulation frequency that a resistance capacitance amplifier will pass without serious attenuation is limited by the stray capacitance across the circuits.

24 RADIO RECEIVER CIRCUITS HANDBOOK

This stray capacitance is the sum total of the small capacitances between the valve electrodes and between the circuit components constituting the circuit and earth. The upper limit is sufficiently high to enable high quality reception of broadcast signals to be provided by a resistance capacitance stage or stages. Attenuation of the modulation frequency voltages at the lower end of the audio-frequency scale is brought about by the impedance of the coupling condenser. At low frequencies the impedance of this condenser becomes high, and the proportion of the total voltage available for passing to the subsequent amplifier stage consequently

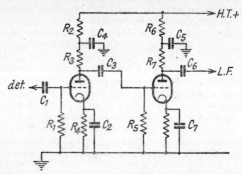

FIG. 12. CIRCUIT OF TWO-STAGE RESISTANCE CAPACITANCE-COUPLED AMPLIFIER

becomes less. It follows, of course, that if the coupling condenser is made large, this drawback does not become effective. On the other hand, too large a coupling condenser brings other difficulties in its train, such as, for example, instability of the amplifier.

In Fig. 12 is shown a circuit that is suitable for resistance capacitance coupling. The input voltage is, in most broadcast receivers, derived from the detector valve and this is applied through condenser C_1 to the control grid of the first valve. The input condenser C_1 may be connected to a tapping, for example, on the diode load resistance of the detector. R_1 is the grid leak of the stage. Correct bias for operating the first valve on the middle of the straight portion of its characteristic curve is obtained by means of the bias resistance R_4 shunted by condenser C_2. This automatic bias arrangement is described in Chapter VIII. At this point it should be mentioned that it is important that R_4 be properly chosen in order that distortionless amplification can take place.

The low-frequency voltages impressed upon the control grid of

LOW-FREQUENCY AMPLIFIER CIRCUITS

the first valve alter the voltage on that electrode both positively and negatively with respect to the normal bias applied by the voltage drop along R_4. These alterations in grid voltage control the electron current flowing from the cathode to the anode of the valve and thereby vary the anode current flowing through the anode circuit. The passage of the direct current from the anode is through resistance R_3, resistance R_2 to h.t. supply, back along the earth wire through R_4 and so to cathode. The variations in the current flow along R_3 set up variations in voltage along that resistance which correspond to the signal voltage in the signal being amplified. These fluctuating voltages are passed through C_3 to the control grid of the second resistance capacitance amplifier. R_5 is the grid leak of the second valve.

It should be noted that quite apart from the use of C_3 for coupling the l.f. voltages from the anode of one valve to the grid of the next, this condenser is necessary in order to block the path of the h.t. voltage. If, for example, C_3 were not present, there would be applied to the control grid of the second valve the full h.t. voltage and the valve would consequently be destroyed. The resistance R_2 is a decoupling resistance and serves, in conjunction with C_4 to avoid any difficulties due to undesired feed back from the h.t. source. This arrangement was described on page 17.

One of the advantages of using resistance capacitance amplifiers is that they are very economical as regards the consumption of the h.t. current. Owing to the use of resistances in the anode circuit, the voltage actually applied to the anode is less than that usually employed with a transformer-coupled amplifier, and consequently the anode current flow is much less. Economy in consumption of anode current, however, should not be considered too important a point, otherwise it will be found that the amplifier will not produce such distortionless reproduction as is desirable. In fact the most satisfactory method of operating a resistance capacitance-coupled amplifier is to use a very much higher h.t. supply than is required by the rating of the valve. This will allow an ample margin for the voltage drop down the anode resistances and still provide a satisfactory voltage at the anodes of the valves. Even so, however, it will be found that the voltage actually at the anodes of resistance-capacitance coupled valves will be much less than that usually applied to the same valves working as transformer-coupled amplifiers.

For the most efficient and satisfactory working of the circuit shown in Fig. 12, the values of R_3 and R_7 should be about three times the value of the anode a.c. resistance of the respective valves. The type of valve most suitable for this circuit is the HL type with an anode a.c. resistance something of the order of 15 000 ohms. In these circumstances R_3 and R_7 could be about 50 000 ohms. The decoupling resistances R_2 and R_6 may be from 10 000 to 20 000 ohms, and the decoupling condensers C_4 and C_5 may be from 1 to 4 μF.

The value of the coupling condenser C_3 is very important, for this determines to a large extent the amplitude of the lower frequencies that are reproduced by the stage as already mentioned. Satisfactory results can be obtained by using 0·02 to 0·05 μF. The grid resistance R_5 should also be chosen with care for this is effectively in shunt to the anode resistance R_3. If it is too low, therefore, the amplification will be reduced. If, however, it is too high, there is a tendency for the second valve to accumulate a negative charge, and in the extreme case the valve may be blocked altogether through this cause. Satisfactory results are obtained by using for R_5 a value of from $\frac{1}{4}$ megohm to 1 megohm, but for preference the lower figure should be tried.

Transformer-coupled Amplifier. In this circuit a low-frequency transformer is employed for transferring the amplified voltages in the output of a valve to the following amplifier or output stage. The valve is biased to the mid-point of the straight portion of its characteristic as already explained. As compared with a resistance capacitance-coupled amplifier, a transformer-coupled amplifier gives a higher amplification owing to the voltage step-up provided by the transformer. Also owing to the fact that as the d.c. resistance of the transformer primary connected in the anode circuit is usually much less than the anode-coupling resistance of a resistance capacitance-coupled stage, the voltage applied to the anode of the valve is much higher. The higher voltage on the anode enables a higher amplification from the valve to be obtained, and also enables a larger grid swing to be handled by the valve without distortion. It is thus seen that a transformer-coupled stage is more sensitive than a resistance capacitance-coupled stage, although the cost of the transformer is usually considerably greater than the cost of the coupling resistance and condenser employed in the latter.

In Fig. 13 is shown a satisfactory circuit for use as a transformer-coupled amplifier. The input coupling condenser C_1, which may be the l.f. coupling condenser associated with a detector valve, applies the l.f. voltages to the grid of the amplifier valve V_1. Suitable bias potential is applied to this grid from the bias battery through grid leak R_1. The signal voltages will vary the anode current of the valve V_1 in a similar manner to that outlined in connection with Fig. 12, and this varying current corresponding in waveform to that of the input signal voltages, passes through the primary winding P of the transformer T.

The alternating current passing round the winding P sets up an electro-magnetic field which induces a voltage in the secondary winding S. The secondary voltage will be amplified relative to that across the primary winding owing to the step-up effect that usually is provided by the greater number of turns in the secondary as compared with those in the primary. Usually the step-up ratio is approximately the same as the turns ratio. That is to say, if the turns ratio of secondary to primary is 3 : 1, then the voltage appearing across the secondary will be about three times that across the primary. It is thus clear that the signal voltage applied to the grid of V_1 is amplified first in the valve itself and

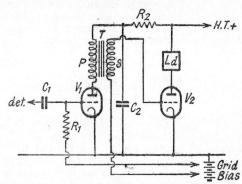

FIG. 13. TRANSFORMER-COUPLED AMPLIFIER

secondly in the transformer. This doubly amplified voltage is passed to the subsequent amplifier valve V_2 as shown.

In practice, the amplification provided by the transformer-coupled stage is not quite so simple of calculation as might appear from the outline given above. For one thing, there is always a certain amount of leakage between primary and secondary windings and as a consequence the step-up ratio provided by the transformer is less than the turns ratio. This is particularly the case when a high turns ratio in the transformer is employed. The higher the turns ratio, the greater is the possibility of leakage. Another and more serious drawback from the point of view of broadcast reception, is the fact that the leakage inductance is a source of distortion for it reduces the amplitude of the higher frequencies in the signal.

Another possible source of distortion is the impedance offered by the primary P in the anode circuit of the valve. Unless this impedance is high compared with the anode a.c. resistance of the valve, the amplification of the stage will not be very great

Now when a low-frequency signal is being amplified, say of 50 cycles, the impedance offered by an inductance such as the primary P, is very much lower than when a signal voltage of higher frequency, say 2 000 cycles, is being amplified. This means that unless special precautions are taken in the design of the transformer T to provide a high impedance primary as compared with the anode a.c. resistance of the valve V_1 even at the lowest desired signal frequency, then the low notes in the music reproduced by the receiver will be either lost altogether or seriously diminished in intensity.

Referring to Fig. 11, it is apparent that the lower drop in the practical amplification curve of a transformer-coupled amplifier is brought about by the diminished impedance of the primary winding at the lower frequencies, and that the drop in the curve at the higher frequencies is due to the leakage inductance of the transformer. Stray capacitances between the windings and between the transformer and earth also tend to reduce still further the amplification of the higher modulation frequencies, and this effect is important in respect of the practical output curve of the amplifier.

In the construction of a low-frequency amplifier it is imperative, therefore, that a suitable transformer T be obtained. A high-turns ratio should not be sought. Normally a ratio of 3 : 1 or 4 : 1 is the most satisfactory from the point of view of quality of reproduced signals. The valve V_1 should have as low an impedance as practicable, consistent with a satisfactory amplification factor so that the impedance of the primary P, which should be as high as practicable, may be several times greater than the anode a.c. resistance of V_1, even at the lower modulation frequencies.

Some care is needed in the adjustment of the grid bias tapping on the bias battery to ensure that the correct operating point is worked upon, and it should be remembered that the anode current of V_1 will be fairly heavy. One result of this comparatively large anode current is that the voltage drop down the decoupling resistance R_2 will be rather high and due allowance must be made for this when working out the required voltage at the anode of V_1. It should also be noticed that the bias applied to V_2 is greater than that applied to V_1. This must necessarily be so, since there is a greater amplified voltage to be handled by V_2, the grid swing being that applied to V_1 multiplied by the voltage amplification of that valve and the transformer.

Parallel Feed Circuit. It will be noticed that in the circuit arrangement of Fig. 13 the entire anode current from V_1 passes through the primary winding of the transformer. Whilst with a robust transformer this passage of current is no disadvantage, it does cause serious distortion when one of the transformers with

LOW-FREQUENCY AMPLIFIER CIRCUITS 29

nickel iron or other high permeability cores is used. When a transformer of this type is employed, the current flow in the primary reduces very considerably the ability of the core to assist the passage of the magnetic field, with the result that the effective inductance of the primary winding is greatly reduced. The result of this is not only a serious reduction in the stage amplification, but also a distortion in the waveform of the signal voltage passed to the subsequent stage. For satisfactory amplification under these circumstances, therefore, it is necessary to use a circuit in which the steady anode current of V_1 is blocked from the path of

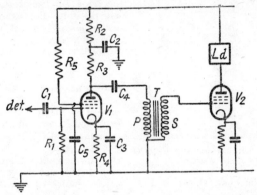

FIG. 14. PARALLEL-FED TRANSFORMER-COUPLED STAGE

the primary winding of the transformer. One such circuit is shown in Fig. 14.

In this circuit the primary of the transformer T is connected to the anode of V_1 via a coupling condenser C_4. The operation of the circuit now is as follows. V_1 acts in a similar manner to the resistance-coupled amplifier shown in Fig. 12. That is to say, the anode current flows along R_3 and an amplified version of the input voltage is developed across R_3. These alternating voltages are passed over coupling condenser C_4 to the primary winding of T, and thence to earth. The alternating currents in P induce voltages in S, and these are applied, of course, to the subsequent amplifier stage V_2 in the usual way.

It is clear from this outline that the steady anode current that is so injurious to the operation of the l.f. transformer T, now passes through the resistances R_3 and R_2, h.t. source, back to the cathode through R_4. The primary of the transformer is in shunt to this circuit and is blocked for direct current by the condenser C_4.

Consequently, only the alternating currents, i.e. the signal currents, pass through the primary P and no harmful effects are thereby produced by the direct current component in the anode circuit.

The only disadvantage to the circuit shown in Fig. 14 as compared with that shown in Fig. 13, is that it costs slightly more to construct. The additional components, however, are only R_3 and C_4, and as against this there is the much higher quality reproduction. The step-up of the transformer arrangement shown in Fig. 14 is

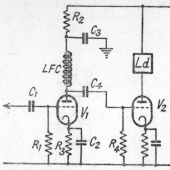

FIG. 15. CHOKE-COUPLED CIRCUIT FOR L.F. AMPLIFIER

precisely the same as that shown in Fig. 13. Other connections of the transformer are possible. For example, it would be quite practicable to join two ends of the transformer (to which the condenser C_4 is connected), the lower end of the primary P being still connected to the earth line but instead of the upper end of S being connected to grid, the lower end would be so connected. This arrangement is using T as an autotransformer and although the step-up is not so great as the arrangement shown in Fig. 14, the quality of reproduction is usually improved.

The size of coupling condenser C_4 has an important bearing on the quality of the reproduction. If it is too low, the lower frequencies are diminished in amplitude. It is frequently advisable to try various sizes of capacitance at C_4 as it is sometimes possible to resonate C_4 and P to one of the lower frequencies in the musical scale. When this is possible, the amplification at the lower end of the musical scale is greatly increased and since, as has already been mentioned, transformer-coupling is rather deficient at this part of the frequency spectrum, the resonating circuit has distinct advantages. It must be realized at the same time, however, that condenser C_4 should be of the high insulation type, for it has to withstand the full value of the voltage applied to the anode of V_1.

Impedance-coupled Amplifier. An impedance-coupled amplifier is one in which a choke is connected in the anode circuit in place of the transformer primary or resistance in the circuits already outlined in this chapter. The arrangement is very similar to the resistance capacitance amplifier of Fig. 12 as a perusal of Fig. 15 will show. Compare also the valve coupling in Fig. 2.

The process of amplification with an impedance-coupled

amplifier is the same as that of a resistance capacitance-coupled amplifier. Instead of the fluctuating anode current producing an alternating voltage along the anode resistance, however, the voltage is produced in the low-frequency choke LFC. For satisfactory results the choke LFC must have a high impedance compared with the anode a.c. resistance of V_1, otherwise the same defect that has been described in connection with low-frequency transformer-coupled amplifiers becomes evident. The alternating voltage is coupled by means of C_4 to the following amplifier valve V_2, the grid leak of this valve being R_4. The stray capacitance in the low-frequency choke forms a limiting factor in respect of the higher modulating frequencies in the signal and gives rise to a falling characteristic at the upper frequencies somewhat similar to the curve shown in Fig. 11.

There is one big advantage of impedance capacitance coupling over resistance capacitance coupling, and this is the much higher voltage that is applied to the anode of the amplifier valve V_1 from a given h.t. source. The d.c. resistance of the choke LFC is usually of the order of 300 to 400 ohms. With an anode current flow of say 8 milliamperes, this gives a voltage drop of only 2·4 to 3·2 volts, which so far as the operation of the amplifier is concerned, is negligible. In the case of resistance capacitance amplification, however, it was seen that the voltage drop along the load resistance was considerable and that this diminished seriously the effective amplification provided by the valve. On the other hand, with impedance-coupled amplifiers there is a risk of undesired resonances in the circuit which tend to spoil the quality of the reproduced signal. There is, however, the possibility of resonating the choke LFC and coupling condenser C_4 at a low frequency in a similar manner to that mentioned with the parallel feed arrangement of Fig. 14. Using a low impedance triode valve as V_1, the impedance of LFC should not be less than 20 henries and should preferably be from 30 to 50 henries.

Impedance-coupled amplifiers are used very little in broadcast receivers. This is because the cost is much greater than that of a resistance capacitance-coupled amplifier and because there is no step-up in the choke like there is in the low frequency transformer.

CHAPTER IV

MULTIPLE VALVE CIRCUITS

IN this chapter are described a number of circuits associated with valves in which more than one cathode-anode stream are employed for separate purposes. Some types of multiple valves are described in Chapter V dealing with frequency changer valves. In the present chapter, however, only those combinations of valves are considered that are used for totally different purposes. For example, detection and automatic volume control are quite separate fundamentally and so are detection and low-frequency amplification. The valves considered in this chapter are mostly of the double-diode-triode and double-diode-pentode types, and the circuits shown are those actually used in receivers and found to be satisfactory in operation.

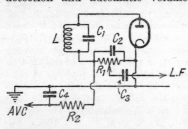

FIG. 16. CIRCUIT FOR PROVIDING A.V.C. VOLTAGES BY DIODE DETECTION

Nearly all the circuits for which double-diode-triodes and pentodes are employed include the production of automatic volume control voltages for application to high-frequency or intermediate-frequency amplifiers. No modern receiver of any standing is without automatic volume control, for it is impossible to listen to many foreign stations without its inclusion in the receiver. In order that the reader may be quite clear about the principles involved in the production of automatic volume control voltages, a description is given very briefly here.

The basic circuit of a diode detector used for combined detection and production of a.v.c. voltage is given in Fig. 16. This circuit is very similar to that shown in Fig. 9 as the basic diode detector, with the addition of a resistance R_2 and C_4. When the incoming signal voltage tuned by LC_1 is applied to the diode, rectification takes place in the manner outlined on page 20, and the rectified voltage appears across diode load R_1. The l.f. voltage is tapped off R_1 and applied to the subsequent amplifier by l.f. coupling condenser C_3, whilst the carrier frequency component

MULTIPLE VALVE CIRCUITS 33

in the diode load is by-passed by condenser C_2. In addition to these voltages there appears along the diode load resistance R_1 a direct current voltage which is dependent upon the actual emission by the cathode and the attraction by the anode. The main factor in this direct current voltage and the one with which we are directly concerned, is the attraction by the anode and this is governed by the signal voltage amplitude. The direct current component is thus proportional to the carrier wave component in the signal, as explained on page 20. As the signal strength increases, therefore, the direct current voltage drop along R_1 becomes greater and, since the voltage at the diode-anode end of R_1 is negative with respect to cathode and to chassis, this may be applied as bias to the control grid of an amplifier valve for varying the amplification.

The d.c. component of voltage along R_1 will not only vary with the carrier wave amplitude but will have impressed on it a low-frequency component which, from the point of view of a.v.c., is not only unnecessary but is definitely injurious. Means have to be provided, therefore, for preventing the passage of any l.f. component to the amplifying valve whose amplification is controlled by the d.c. voltage drop along R_1. Resistance R_2 and condenser C_4 are provided for smoothing out any l.f. or h.f. components, and these are known as the filter elements. A common value for R_2 is a megohm, but a fair amount of latitude is permissible so long as its value is not made too low. The filter condenser C_4 is usually about 0·1 μF. This condenser should not be too low in capacitance, otherwise its effectiveness as a smoothing condenser may be seriously diminished.

The basic circuit given above may be modified very considerably in practice, as will be seen later, but its description will help to clarify any doubts as to precisely how the a.v.c. systems to be described actually work. This will also avoid the necessity of explaining this part of the circuit in each figure described.

Simple A.V.C. and L.F. Amplification. A commonly used double-diode-triode circuit is given in Fig. 17. The tuned input circuit is LC, which in most circuits will be the tuned i.f. secondary of the final i.f. amplifier. I.F. voltages are applied to the two diodes which are connected together. This arrangement is known as strapped diode detection, and merely consists in using the diodes in parallel. The diodes, therefore, have a common load resistance which is R_2. R_1 helps to filter out the i.f. component from the diode load, and it is noticed that two shunt condensers C_1 and C_2 are used to help this filtration.

From off the diode load R_2 is tapped a portion of the l.f. component and applied by a coupling condenser C_3 to a fixed resistance R_3. R_3 is, of course, the grid resistance of the triode portion of the valve and the upper end of it is connected to the grid. The rectified voltages are thus applied to the grid of the triode and amplified in the usual manner by the triode portion as though the diodes were not present in the valve. In other words, the operation of diodes in a multiple valve may be considered from a practical point of view, with which this book is concerned, as being quite apart and separated from the operation of the triode or other portion of the valve. In Fig. 17 the amplified voltages in the anode circuit and appearing across the coupling

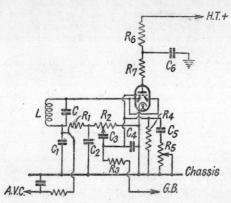

Fig. 17. Double-diode-triode Circuit for Detection and L.F. Amplification

resistance R_7, are applied by a coupling condenser to the following amplifier, which usually is an output pentode.

The resistance R_2 acts as a manual volume control, for by means of the tapping on this resistance any desired proportion of the total a.c. voltage along it may be applied to the grid of the triode. For maximum output the tapping point along R_2 must be that farthest from the cathode lead. Additional filtration of any residual i.f. component is effected by condenser C_4, whilst a tone control is provided by the combination of C_5 and R_5. As the value of R_5 that is inserted in circuit increases, so the effectiveness of C_5 as a shunt across the grid leak R_3 becomes less and the reproduction becomes more high pitched. Although the arrangement shown between R_3 and R_5 seems a little complicated, it is seen that each component has a very distinct purpose. Condenser C_6 is to decouple the anode circuit and works in conjunction with the decoupling resistance R_6.

Practical values of the components for this circuit vary somewhat according to the type of valve used, but the following will be found to represent very useful values: R_1—50 000 ohms, R_2—500 000 ohms, C_1—0·0001 μF., C_2—0·0001 μF. C_3 has the usual l.f. coupling condenser value of 0·02 to 0·05 μF., and grid bias resistance R_3 may be about $\frac{1}{2}$ megohm. C_4 should not be more than about 0·00025 μF., otherwise the reproduction will become low pitched. R_4 should be round about $\frac{1}{4}$ megohm and if C_5 is given a value of 0·01 μF. very effective control of tone will be obtained with a resistance R_5 of $\frac{1}{2}$ megohm. The anode resistance R_7 should be about three to five times the value of the anode a.c. resistance of the triode portion and the decoupling resistance R_6 should have the usual value of about 10 000 to 20 000 ohms. Decoupling condenser C_6 should be from 2 to 8 μF. Decoupling by R_6C_6 is not always necessary, although often advisable.

Double-Diode-Pentode : Delayed A.V.C. and L.F. Amplification.

Most of the circuits for which the double-diode-triode and double-diode-pentode are employed can be used for either of these types of multiple valves, so long as suitable modifications, such as, for example, the anode resistance or inductance values, are made to suit both cases.

Simple a.v.c. as used in the circuit of Fig. 17, although quite a common circuit for use in connection with double-diode-triodes and double-diode-pentodes, has certain disadvantages. It is frequently desired that the a.v.c. system shall not operate until the sensitivity of the receiver has reached a level at which the reproduction is sufficiently loud to have good entertainment value. Under these circumstances, a method of a.v.c. known as delayed a.v.c. is used.

The principle underlying this method of control is to block the a.v.c. rectifier until the signal voltage applied to it has reached a certain value. Up to this point the receiver acts as though there were no automatic volume control whatever, and operates at full sensitivity. As soon as the value of signal voltage required to operate the a.v.c. device has been attained, the control operates and reduces the sensitivity of the receiver in accordance with the strength of the incoming signal.

The usual method of delaying the action of the a.v.c. is to apply a negative voltage to the a.v.c. diode-anode. It has already been shown that, in order that the diode shall pass current, there must be a positive voltage at the anode. If, therefore, the diode-anode is made negative by say, four volts, then the diode cannot detect incoming signals until their peak value exceeds four volts.

A satisfactory and practical circuit for providing delayed a.v.c., second detection and low-frequency amplification in a double-diode-pentode is given in Fig. 18. Tuned circuit L_1 C_1 which may be, of course, the tuned secondary winding of the final i.f. transformer, applies voltage to the signal diode D_1. Signal diode load resistance is R_1 and this is by-passed for the carrier frequency component by condenser C_3. The choke is for the

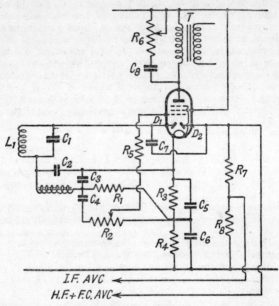

Fig. 18. Double-diode-pentode Circuit in which One Diode is Used for Detection and the Other for Delayed A.V.C.

The pentode is an L.F. amplifier

purpose of preventing the carrier-frequency component from passing into the low-frequency part of the circuit and works in conjunction with condenser C_2 to block this path. The operation of the signal detector circuit is a normal one.

The a.v.c diode is D_2 and it is noted that this diode is connected by means of a small condenser C_7 to the signal diode anode D_1. Voltages at carrier frequency are thus applied simultaneously to both diodes, and the a.v.c. diode D_2, therefore, rectifies the signal voltages in a similar manner to D_1. The point to be noted, however, is that D_1 and D_2 are not in parallel in respect of d.c.

MULTIPLE VALVE CIRCUITS

as they are in Fig. 17, owing to the position of C_7. It is possible, for example, to apply to D_2 a negative d.c. voltage for purposes of delaying the action of the a.v.c. and yet not to affect the operation of D_1 whatsoever. This delay voltage is provided by the resistances R_3 and R_4 which, it should be noted, are connected in the cathode lead of the valve. When the operating voltages are applied to the valve, the normal anode current flows between cathode and pentode anode, and this sets up the usual voltage drop in the resistances R_3 and R_4. As the diode-anode D_2 is connected through its load resistances R_7 and R_8 to the earth line, the anode of D_2 must be negative with respect to the cathode to the extent of the drops down R_3 and R_4. Consequently the action of D_2 is delayed.

The total voltage drop in the load resistances of the a.v.c. diode D_2 is not applied to all the previous valves in the receiver, but instead it is split up into two separate parts. The full voltage, i.e. that appearing across R_7 and R_8, is applied to the high-frequency amplifier and to the frequency changer. That portion appearing across R_8 is applied to the i.f. amplifiers only. By this means it can be arranged that there is more effective control on the earlier stages of the receiver than on the later stages and this usually brings about a more satisfactory method of control than if the same voltage is applied to all the valves required to be controlled.

The return lead of the signal diode D_1 is not taken directly to the cathode, but is joined to a point intermediate resistances R_3 and R_4, providing positive voltage on the cathode for purposes of delayed a.v.c. and grid bias. The result of this is that the anode of D_1 is made slightly negative to the cathode. The amount of this negative bias of D_1 is that provided by the voltage drop along R_3. The signal diode is, therefore, rendered inoperative until the signal voltage applied to it by the tuned circuit $L_1 C_1$ exceeds this bias. This arrangement is known as a simple noise suppression circuit, for by its use the reproducer is rendered silent when the receiver is made most sensitive, i.e. when very weak signals or no signals are being received. It is not an effective noise suppression circuit or quiet a.v.c., but it does serve the useful purpose of making the receiver less noisy when tuning round the dial, than if a negative bias is not applied to the signal diode at all.

The output circuit consists of an output transformer T, the primary of which is shunted by the usual tone corrector circuit R_6 and C_8. In this instance the screen grid is connected directly to the h.t. line. The values of the components are similar to those mentioned in connection with Fig. 17. The additional circuit elements are the stopper resistance R_5, which may be from 15 000 to 100 000 ohms, and the a.v.c. diode load resistances R_7 and R_8. These should be dimensioned in such a way that the requisite voltages are applied to the respective valves of the receiver. It sometimes happens that the control voltage required to be applied to the frequency changer and high-frequency valve is nearly three times that applied to the intermediate-frequency amplifier. In this case R_7 should be about double the value of R_8 and, although the values will be dependent upon the type of a.v.c. diode employed, generally satisfactory values are $\frac{1}{2}$ megohm for R_7 and $\frac{1}{4}$ megohm for R_8.

The dimensions of the grid bias resistance R_3 must be calculated according to the grid bias required by the pentode portion of the valve, and also it must work in conjunction with R_4 to provide the delay voltage for the a.v.c. diode. At the same time it should be remembered that R_3 also has to provide the negative voltage for the signal diode in order to render this valve operative as a noise suppression device between stations. It may be found that if R_3 is made double the value of R_4, these conditions are satisfactorily fulfilled, but final adjustments should be made after the circuit has been in operation. The by-pass condensers C_5 and C_6 should have a large capacitance in view of the fact that low-frequency voltages are being dealt with by the pentode. A capacitance of 25 μF. for each will not be excessive and electrolytic condensers may be used if required, the positive terminals being connected to the cathode side. The tone corrector across the transformer primary has the usual values of 25 000 ohms for the resistance and 0·02 μF. for the condenser.

Alternative Circuit for Double-diode-pentode.

Owing to the popularity of the double-diode-pentode, an alternative circuit is given in Fig. 19 for its operation as a second detector, a.v.c. valve and l.f. amplifier. This circuit is not very dissimilar to that given in Fig. 18, but it may help the reader to form a better idea of the capabilities of this valve, and will in any event, serve as a guide to show the type of circuits that can be employed with this valve.

The signal diode is D_1 and the a.v.c. diode is D_2 as in the previous instance. Signal diode load is R_1, and this is by-passed for the carrier frequency by C_2. The l.f. shunt is C_3 and R_2, and the signal diode return goes to cathode. It should be mentioned that this arrangement is different to that given in Fig. 18. In the present case, diode D_1 is at cathode potential and no noise suppression is obtained. This circuit is, nevertheless, very satisfactory and is commonly used. The arrangement shown here is more simple than that given in Fig. 18, and less components are required. For this reason, the circuit shown in Fig. 19 might appeal to many designers that would not be particularly keen on the arrangement shown in Fig. 18.

The a.v.c. diode D_2 is connected by a condenser to the anode of the last intermediate-frequency valve by means of the small condenser C_5. It should be mentioned that the method of tapping off the a.v.c. carrier voltage is not of vital importance to the operation of the receiver. Both the arrangements shown in Figs. 18 and 19 are in common use. Taking the a.v.c. voltage from the i.f. anode has the advantage of providing a higher a.v.c. voltage than when the a.v.c. carrier voltage is shunted away from the second detector circuit as shown in Fig. 18. However, there are pros and cons to both methods.

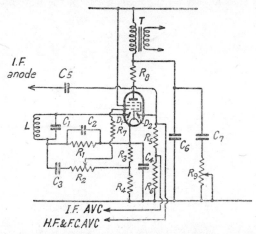

Fig. 19. Alternative Double-diode-pentode Circuit for Detection, A.V.C., and L.F. Amplification

To the anode of the pentode portion is connected a resistance R_8 which is designed to stop the outbreak of parasitic oscillations. A normal value for this resistance is from 50 to 150 ohms. It should not be greater than is necessary, otherwise the circuit will not work very well.

The corrector circuits for the pentode are shown across the valve between anode circuit and earth, the value of C_6 is about 0.0001 μF. and the corrector C_7 and R_9 can be given values of 0.02 μF. and 50 000 ohms respectively.

The other parts of the circuit are very similar to those shown in Fig. 18 and will not be further commented upon here.

Double-diode-pentode as High-frequency Amplifier, Second Detector, and A.V.C. Valve. The circuits for use with multiple valves so far given in this chapter appear rather complicated. Although they will be found very satisfactory, there may be a requirement for a more simple arrangement which will also be satisfactory although not perhaps quite up to the standard of the somewhat more complicated circuits. For those who require the simple circuit, the arrangement shown in Fig. 20 is given.

In this arrangement, the pentode portion is used as a high-frequency amplifier and the amplified high-frequency voltages are fed back to the diodes which then function as a normal detector and a.v.c. valve. The signal voltages are applied to $L_1 C_1$ and the voltages thus appear on the control grid of the pentode. In the

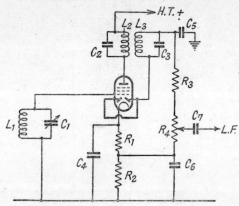

Fig. 20. The Pentode Portion is here used as H.F. or I.F. Amplifier, the Diodes being in Parallel for Detection and A.V.C.

output is the tuned circuit $L_2 C_2$ coupled to L_3, also tuned by C_3. This arrangement is the usual doubly tuned transformer and the signal voltages appearing across the secondary tuned circuit are applied to the two diodes which are strapped together. Rectified voltage thus appears across the diode load resistance R_4 and is tapped off by C_7 for application to the low-frequency amplifier. The arrangement of biasing resistances R_1 and R_2 is similar to that already outlined in connection with previous diagrams, and a certain amount of inter-station noise suppression is provided by the connection from R_4 to the mid-point of R_1 and R_2. C_5 is the usual by-pass condenser.

The values of the components used in Fig. 20 are similar to those already described in connection with the previous circuits.

Triode-pentode as Intermediate-frequency Amplifier and Audio-frequency Amplifier. In Chapter V is described the use of the triode-pentode as a frequency changer. Although this valve was originally designed as a frequency changer for superheterodyne receivers, it can act very satisfactorily as an amplifier of intermediate-frequency and audio-frequency voltages. The pentode portion, for example, is an admirable amplifier of the intermediate frequencies, while the triode portion can operate as an amplifier of low-frequency voltages. In commercial receivers, in fact, the triode-pentode is sometimes employed for the dual purposes mentioned above. The advantage of using the valve in this manner

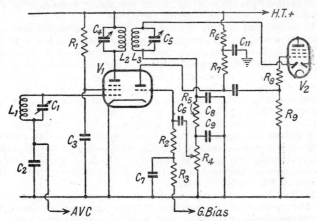

FIG. 21. THE PENTODE PORTION OF THE TRIODE-PENTODE V_1 IS HERE USED AS H.F. OR I.F. AMPLIFIER, A DIODE OF V_2 AS DETECTOR, THE TRIODE OF V_1 AS L.F. AMPLIFIER AND THE PENTODE OF V_2 AS OUTPUT VALVE

is mainly one of compactness, for it is a common desire to get the utmost amplification from a given number of valves regardless of whether these so-called valves have one or more sets of electrodes within the envelope. For this reason the triode-pentode has much to commend it as providing within one envelope an arrangement capable of affording a very high voltage gain.

One circuit for using the triode-pentode for amplifying i.f. and l.f. voltages is shown in Fig. 21. The i.f. signals are passed by $L_1 C_1$ to the control grid of the pentode and are amplified at the i.f. frequency in the usual way. In the output of this valve there is the primary winding L_2 of the i.f. transformer, the secondary winding L_3 being tuned by C_5. The i.f. voltages are thence passed to the signal diode of the following amplifier V_2 and the rectified voltages appear across the diode load resistance R_4. From R_4 there is tapped a low-frequency voltage and this is applied through coupling condenser C_6 to the grid of the l.f. amplifier portion of V_1, i.e. the triode. Resistance R_2 is the normal grid leak of the triode amplifier of V_1 and R_3 is the grid decoupling resistance which works in conjunction with decoupling condenser C_7. Amplified l.f. voltages are produced in the anode circuit of the triode, and are coupled by a condenser to the grid of the following l.f. amplifier shown in the diagram as the pentode portion of a double-diode-pentode valve. R_7 is the anode resistance of the triode and R_6 is the de-coupling resistance.

It is seen that in the circuit of Fig. 21 a very compact arrangement is provided. I.F. amplification takes place in the pentode of V_1, detection by the signal diode, l.f. amplification in the triode of V_1, and then further amplification in the output pentode of V_2. Assuming that the remaining diode of V_2 is used for providing a.v.c. voltages, it is thus seen that in two valves, V_1 and V_2, there are five distinct functions, i.e. i.f. amplification, second detection, l.f. amplification, power amplification and a.v.c. Although the same number of components is required whether two valves as shown in Fig. 21 are employed or five valves, there is a definite economy in initial cost owing to the necessity for purchasing two valves as against five, and also an economy in heater current required from the mains transformer when the circuit of Fig. 21 is employed.

Most of the components used in Fig. 21 are similar to those used in the corresponding portions of other circuits. It may be worth mentioning that the grid leak R_2 may have a value of about $\frac{1}{2}$ megohm, stopper resistance R_5—50 000 ohms, diode load resistance R_4—$\frac{1}{2}$ megohm, and the anode resistance R_7 from 50 000 to 100 000 ohms, while grid-leak resistance R_9 may have a value of about $\frac{1}{2}$ megohm. All the values, however, should be a matter of experiment, but the figures given will serve as a guide.

In operating a circuit of this type, it is important to ensure that the grid bias supplies for V_1 are separate, owing to the two valves therein having a common cathode. In Fig. 21 the pentode portion is biased by the a.v.c., whilst the triode is preferably biased by the voltage drop along an independent resistance. Decoupling for the triode grid bias is provided by R_3 (about $\frac{1}{2}$ megohm) and C_7 (about $0 \cdot 1$ μF.).

CHAPTER V

FREQUENCY CHANGER CIRCUITS

THE process known as frequency changing in radio receivers is that which takes place in a superheterodyne receiver for reducing the frequency of the carrier wave from the incoming signal frequency to that known as the intermediate frequency. When an incoming signal consisting of carrier wave and sidebands is received and amplified, various difficulties occur when it is desired to provide a high degree of amplification. For instance, most high-frequency amplifier valves, both screen-grid valves and pentodes, are inclined to be unstable when they are worked in such a way that they provide a high stage gain. This was explained in Chapter I in connection with high-frequency amplifiers.

Owing to the limitations that are imposed on the straightforward high-frequency amplifier, it is found desirable when a high overall gain is required, to change the frequency of the carrier to a lower one and thus enable a stable amplifier of high gain to be obtained. The process whereby this change of frequency takes place is known as frequency changing and the valve in which the process is carried out is the frequency changer valve.

The advantages of changing the frequency from the signal frequency to the intermediate frequency comprise not only the provision of a much more stable amplifier and thereby an amplifier that may include a larger number of amplifier stages and consequently provide a higher overall gain, but also the construction of a receiver to give a much higher selectivity. This increased selectivity is obtained partly by the use of the lower frequency itself and partly due to the facility with which the intermediate-frequency amplifier coils and tuned circuits can be constructed. It is an easily demonstrable fact that circuits tuned to a low frequency are much more inherently selective than circuits tuned to a higher frequency. This increased selectivity is partly due to the simple arithmetical fact that for a given frequency difference an interfering station will be a greater percentage out of tune when the desired station is at a lower frequency than if the desired station has a higher carrier frequency. For example, if a desired station has a carrier wave of 1 000 kc., then the station on the adjacent carrier wave will have a frequency of 1 009 kc. The difference in frequency between the desired and the interfering signals is, therefore, 9/1 000, or something like 1 per cent. It is difficult to provide a straightforward amplifier that will separate an interfering station so close to the desired station if that interfering station is strong in comparison to the desired station. If now the carrier frequency of the desired station is transformed down to 100 kc., then the frequency of the interfering station, which will, of course, have been transformed correspondingly, will still be separated from the desired signal by 9 kc., but the percentage off tune is now 9/100, or 9 per cent. It is a much more simple proposition to construct a tuned circuit to separate stations differing by 9 per cent in frequency than in the former case differing by only 1 per cent.

Having made out the case for frequency changing, and, in doing so, for the superheterodyne receiver, it should be mentioned that difficulties are involved in this process of frequency changing. For instance, it is not an easy matter to construct a frequency changer stage that will work satisfactorily, owing to the presence of stray capacitances between the circuits and to a certain amount of interaction that frequently results from the use of one valve for this process. There are a large number of other factors that enter into this subject, but they are not required to be dealt with here. These preliminary notes are merely an introduction to the process of frequency changing. For further information on frequency changing and the superheterodyne receiver, the reader is referred to *The Superheterodyne Receiver*, by A. T. Witts (Pitman).

The circuits outlined in this chapter are those that are used in modern superheterodyne receivers and have been found to be entirely satisfactory if used in the manner outlined. In some of the circuits, difficulties are encountered if an effort is made to use the valve at too high a frequency, but it may be a matter of experiment by the reader to find out just how high that frequency really is. On the other hand, some of the valves described below have been particularly designed to work on very high frequencies, and these are mentioned in the relative descriptive matter.

Heptode Frequency Changer. A heptode is a valve that has seven electrodes, five of which are grids. It is usually understood that the five grids are in the space between the cathode and the anode. All grids thus have an influence on the electron stream flowing from cathode to the main anode. There are, of course, a number of constructions of valve that comprise seven electrodes, but unless the valve has the five grids in the position mentioned here, the valve is not normally referred to as a heptode.

A heptode circuit is given in Fig. 22. The inductance coils L_1 and L_2 for medium and long waves respectively form the secondary winding of an input h.f. transformer or coupling coil and are fed by either the aerial circuit or a signal-frequency amplifier. The input circuit to the heptode is thus comprised of one or both of the coils L_1 L_2 tuned by variable condenser C_1. This circuit is tuned, of course, to the carrier wave of the desired incoming signal and voltages of this frequency are applied to the signal grid which is the fourth grid counting from the cathode. It should be mentioned here that in multi-grid valves, the grids are usually counted from the cathode in the direction of the anode for reference purposes.

The local oscillation circuit consists of inductances L_6 and L_7 which are tuned by the variable condenser. The second grid is

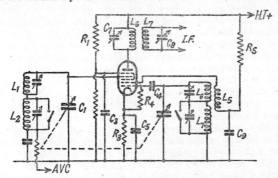

Fig. 22. A Heptode Frequency Changer Circuit

connected to a coil L_5 and thence to the h.t. source. The first grid is known as the oscillator grid and the second grid as the oscillator anode. This is because the cathode, first grid, and second grid are operated in such a manner that they form an oscillation generator and may be considered as serving the purpose of producing the oscillations required for frequency changing and do not take a part in the actual amplification of the signal.

Owing to the feed back of voltage that takes place from L_5 to L_6 and L_7, the voltage at the oscillator grid is augmented until oscillations take place and then a state of equilibrium is arrived at in which oscillations are maintained at a reasonably constant amplitude and frequency. The actual frequency of these oscillations is determined, of course, by the frequency of the tuned grid circuit consisting of L_6 and L_7 together and the tuning condenser. Voltages at this oscillation frequency are applied to the oscillator grid via grid condenser C_4.

The grid itself is connected to cathode through grid leak R_4. The function of C_4 and R_4 should not be confused with that of the grid condenser and leak of the grid detector. In the present case they serve as a means of stabilizing the actual oscillator voltage at the grid owing to the voltage drop produced down R_4 when grid current flows. This voltage drop places a negative bias on the oscillator grid, and, should the amplitude of the oscillations generated tend to become too high, the negative bias applied increases and this reduces the electron current reaching the oscillator anode, and as a consequence the feed back is diminished. This reduces the amplitude of the voltage across the tuned grid circuit and thus the oscillator voltage is levelled up.

The grids 3 and 5, situated on either side of the signal grid 4, shield the signal grid and circuit from the effects of the oscillator voltage. It has already been mentioned in the preliminary notes in this chapter that one of the difficulties with frequency changer valves is the tendency towards interaction between the circuits. Grids 3 and 5 serve to diminish the interaction and thus to isolate, as it were, inside the valve the electrodes on which voltages at the three different frequencies appear. The screen grids are held at a positive d.c. potential which is usually, but not always, lower than that of the oscillator anode. It is essential that a decoupling condenser be connected between these screen grids and earth or cathode by a non-inductive decoupling condenser. If this is done, any h.f. currents at the frequencies being handled by the valve that reach the screen grids will be passed immediately to earth. Under these conditions, the screen grids form effective screens, as explained more fully on page 3.

Connected to the main anode of the heptode is a transformer, the windings of which are tuned to the intermediate frequency by C_7 and C_8. The coupling between the windings is usually adjusted to give a band pass filter effect.

The complete process will now be apparent. Local oscillations are generated by the first two grids and as a result the electron stream passing through the oscillator anode towards the main anode is modulated at the voltage of the oscillations produced. The electron stream is further modulated as it passes the fourth or signal grid, this time at the voltage of the incoming signal frequency. The combination of these two modulations is such that there are produced in the anode circuit voltages at a number of frequencies, one of which is equal to the difference between the oscillator frequency and the signal frequency. This difference frequency is the required intermediate frequency and is selected by the tuned transformer in the anode circuit.

In general, the heptode valve is a very easy one to operate, and is not likely to give much trouble if the voltages recommended by the valve manufacturer are employed. It is not advisable to use very high screen voltages that are frequently employed in low-frequency amplifier valves, as this is liable to cause the valve to

FREQUENCY CHANGER CIRCUITS

oscillate as a unit and render reception impossible. The value of grid leak R_4 may vary between 50 000 ohms and 250 000 ohms, but the former figure is usually preferred for the most satisfactory operation. Grid condenser C_4 is usually of the order of 0·0001 μF., but this value will depend to some extent on the actual type of valve employed, in the same way that the value of the grid leak does.

Grid bias resistance R_3 is the minimum grid bias resistance required to prevent the valve being operated at zero grid bias, for it is noted that a.v.c. is used with this valve. A.V.C. is invariably employed in superheterodyne receivers owing to the high overall gain that this type of receiver provides. The minimum bias resistance R_3 is generally about 250 ohms, but it may be less, of course, if the valve is suitable. On the other hand, R_3 should not be reduced more than is really necessary, for too low a value may bring about undesired oscillation of the entire stage. If there is any doubt about the value of R_3, 500 ohms should be used for a start and reduced if desired. By-pass condenser C_5 must, of course, be of the non-inductive type and from 0·1 to 0·25 μF. capacitance.

The screen-grid potentiometer is not always necessary. In fact, quite a large number of valves merely require a voltage dropping resistance between the h.t. lead and the screen grid itself, the second resistance in these cases being superfluous. The two resistances are shown in Fig. 22 for completeness only. The screen-grid condenser C_3 must also be of the non-inductive type as this performs a similar function to the normal screen-grid decoupling condenser referred to on page 3.

In practical frequency changer circuits using the heptode, not many modifications of the circuit shown in Fig. 22 are employed, as this circuit is found to be entirely satisfactory. An alternative position for the grid leak R_4 is across the grid condenser C_4, instead of between the oscillator grid and cathode. When the grid leak is in this position, the oscillator grid is given the bias provided by the grid bias resistance R_3 and this should be remembered when the shunt arrangement of the grid condenser and leak is employed.

An alternative position for the grid leak and condenser is in the oscillator grid return to cathode. The operation of the valve is then similar to that when the grid leak and condenser are connected in the position mentioned in the preceding paragraph, except that the oscillator grid is not biased by the resistance R_3.

Octode. The octode is a valve with eight electrodes, six of which are grids in the electron path between cathode and main anode. The arrangement of the circuit is similar to that used with the heptode and the foregoing notes are applicable to the octode circuit. Octodes generally have a higher anode a.c. resistance than heptodes, and for this reason give a somewhat higher order of selectivity. The extra grid as compared with the heptode is connected between the screen grids and main anode in the electron stream. This extra grid is connected to either the cathode or to earth in a similar way to the third or suppressor grid of a pentode valve.

Triode-pentode. This valve consists of a triode portion serving as local oscillator and a pentode portion that is used as the mixer

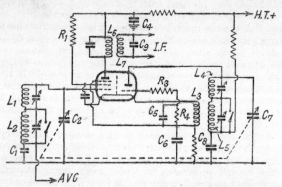

FIG. 23. THE TRIODE-PENTODE AS FREQUENCY CHANGER

of the incoming and the locally produced oscillations. The two sets of electrodes are screened from each other inside the valve, but employ a common cathode. The valve has been designed to provide a high degree of stability of local oscillations and freedom from harmonics, and to prevent as far as possible the interaction of the two sets of oscillations. On page 97 is described the more modern electron coupled triode pentode.

The basic circuit for use with this valve is given in Fig. 23. A tuned input circuit consisting of L_1 (medium waves) and L_2 (long waves) in conjunction with variable condenser C_2 applies voltages of the incoming carrier frequency to the control grid of the pentode portion. The local oscillations are produced by the triode portion and determined by the resonant frequency of L_4 and L_5 in combination with variable condenser C_7. It will be noted that the tuned circuit of the oscillator is the anode circuit, whereas in the heptode arrangement shown in Fig. 22, the tuned circuit is the grid circuit. Voltages of the local oscillation frequency are injected into the cathode circuit by means of the coupling between L_3 and L_4/L_5. As the actual feed back circuit

FREQUENCY CHANGER CIRCUITS 49

is in the cathode lead, this arrangement is known as cathode injection. The cathode is thus not at earth potential, but has a potential that is varied by the local oscillations.

The injected oscillatory voltage in L_3 is applied simultaneously to the pentode portion owing to the cathode being common to both triode and pentode. Owing to the influence on the electron stream of the pentode, frequency mixing takes place in this valve. The difference or intermediate frequency is selected by the primary L_6 of the intermediate transformer consisting of L_6 and L_7, both of which are tuned to resonance at the intermediate frequency by preset condensers.

It should be noted that the grid bias resistance shunted by C_6 is not effective for the oscillator valve, since the grid of the oscillator triode is returned to the cathode end of this resistance. Consequently the oscillator grid will be at a negative voltage determined only by the grid current flow down R_3 and R_4. R_4 is the usual grid resistance and C_5 the grid condenser, while R_3 is a harmonic suppressor and serves the purpose of limiting the voltage of spurious harmonics that are usually present in all oscillator circuits. The grid bias resistance provides a minimum bias for the control grid of the pentode mixer portion of the valve.

It will be noted that in this circuit the screen grid of the pentode is not supplied with voltage from a potentiometer as shown in Fig. 22, but is connected to h.t. supply by one resistance only, R_1. The screen grid decoupling condenser is returned to cathode. If this condenser is joined to earth, feed back occurs at the frequency of the oscillator circuit. It is also preferable to decouple the pentode anode to cathode.

The grid bias resistance has a value dependent upon the type of valve, and should not be too low if instability is to be avoided. Resistance R_4 is commonly about 40 000 ohms and C_5 is 0·0001 μF. The harmonic suppressor R_3 is generally of the order of 2 000 ohms.

The triode-pentode is not used so much in modern receivers as it was three or four years ago. It is, nevertheless, employed owing to the simple circuit that may be used with it. The triode-hexode has tended to oust the triode-pentode owing to the shorter wavelengths that may be received by the former.

The triode-pentode may be used with oscillator and input circuits completely isolated, the oscillator voltage being applied to the suppressor grid by a condenser (0·0005 μF.) connected to the triode anode. The suppressor grid is connected to earth by a leak (2 megohms).

Triode-Hexode. The hexode is a valve with six electrodes and includes four grids in the direct electron path between cathode and anode. Hexodes were used to a considerable extent on the

Continent before they were introduced in this country, but they have become very popular in the combination of triode and hexode as a frequency changer owing to the special construction that forms a satisfactory frequency changer on very high frequencies. Satisfactory reception of the short and ultra-short wavelengths is thereby facilitated.

At first sight, the combination of triode and hexode would appear to be similar to that of a triode and pentode. Actually,

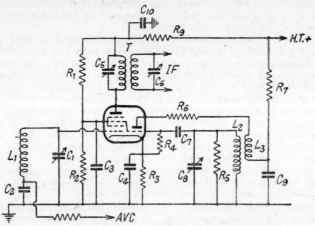

Fig. 24. Circuit of the Triode-hexode Frequency Changer

however, the circuit arrangement is quite different, for with the triode-hexode, the coupling between the two portions of the valve is by a direct connection between two of the grids, whereas in the triode-pentode circuit of Fig. 23 the coupling is by means of a coil in the common cathode lead. The arrangement of the coupling in the case of the triode-hexode is such that the interaction between the voltages of the two frequencies necessary for frequency changing is reduced to the minimum.

A commonly used circuit for the triode-hexode is shown in Fig. 24. The input circuit L_1 C_1 applies voltages of the incoming signal frequency to the control grid of the hexode. It will be noted that the control grid is in the same position as in the pentode, i.e. is the first grid counting from the cathode. Oscillations are generated by the triode portion by virtue of feed back between the anode coil L_3 and the grid coil L_2. The oscillation determining circuit is L_2 C the anode coil being aperiodic. A grid condenser

C_7 and grid leak R_4 are used for stabilizing the oscillation amplitude.

The triode oscillator portion is seen to be entirely separate from the hexode portion, not being connected in any way whatever, except through the common connection shown between the triode grid and the third grid of the hexode. This connection is inside the valve, and by means of it voltages at the oscillator frequency are applied directly to the electron stream flowing between the cathode and anode of the hexode valve. This is what is known as electron coupling. This type of coupling makes external coupling coils and condensers quite unnecessary. The arrangement in fact is similar to that described in connection with the heptode, with the exception that instead of the oscillator electrodes being in the common electron stream and modulating this, the oscillator electrodes are now quite separate but modulate the mixer valve electron stream by means of the connection between the grids as shown. The employment of a separate triode in this way has the advantage, as compared with the heptode, of enabling a triode with a higher mutual conductance to be constructed and, in this way, of obtaining an oscillator which will operate satisfactorily on the lower wavelengths where the heptode is likely to cause difficulty owing to the lower value of mutual conductance between the oscillator electrodes of that valve.

The intermediate frequency is selected in the usual manner by tuned i.f. transformer T, the windings of which are tuned by C_5 and C_6 respectively. Screen-grid potentiometer $R_1 R_2$ is not always necessary, R_1 alone being used with some valves. R_6 is a voltage limiter and serves to prevent the oscillator voltage rising to too high a value and thereby spoiling the operation of the valve. R_5 is a damping resistance for preventing an uneven oscillatory voltage and operates in such a way that, on the wavelengths at which the oscillator works most efficiently, the effect of the damping resistance R_5 is greatest. This reduces the oscillator voltage applied to the hexode portion and thus tends to level the oscillator or heterodyne voltage generated at the various wavelengths.

Grid leak R_4 may have the usual value of about 50 000 ohms and C_7 a value of about 0.0001 μF. R_6 is dependent upon the waveband and type of valve used, but values of from 100 to 2 000 ohms are commonly used. The damping resistance R_5 may not be necessary, but, if used, should have a value of from 10 000 to 50 000 ohms, depending entirely on the operation of the valve, best found by experiment.

One advantage of the triode-hexode as compared with the heptode is that when a strong signal is being received, the total anode current is reduced owing to the action of the a.v.c. on the control grid of the hexode portion. The anode current of the triode portion remains sensibly constant, of course. Now with a heptode or octode, the total current flowing from cathode to the various electrodes is about the same, irrespective of the voltage applied to the signal grid. Although, when mains valves are being used, there is not much object in economy in anode current during the reception of strong signals, when dry batteries are used as the source of h.t., any economy in this respect is welcome. The triode-hexode, therefore, scores over the heptode from the point

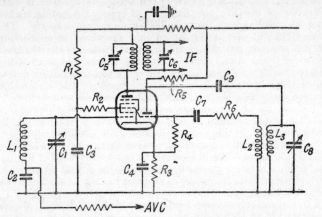

FIG. 25. THE TRIODE-HEPTODE CIRCUIT

of view of consumption of h.t. battery when strong signals are being received.

Triode Heptode. A circuit for using the triode-heptode is given in Fig. 25. The valve used here has a triode oscillator combined with a valve having five grids. The latter has seven electrodes, and the combined valve is called a triode-heptode.

The main difference between Fig. 24 and Fig. 25 is in the connections of the triode oscillator. In Fig. 25 the tuned oscillator circuit $L_3 C_8$ (determining the oscillator frequency) is joined to the anode instead of to the grid of the triode. Energy for the tuned oscillatory circuit is derived from the triode-anode via coupling condenser C_9. D.C. voltage is applied to the triode-anode by the connection of R_5 (35 000 ohms) to the h.t. supply. This arrangement for feeding the tuned anode circuit is known as parallel feed, as the circuit $L_3 C_8$ is parallel to the feed circuit consisting of h.t. supply and R_5.

It is noted that the triode grid circuit is untuned. There is no

FREQUENCY CHANGER CIRCUITS

need to tune both grid and anode circuits of an oscillator, and a tuning condenser across L_2 would be superfluous in ordinary receiver circuits. R_6 is to stabilize the oscillator voltage and should have a similar value to that mentioned in connection with Fig. 20. Grid condenser C_7 has the usual 0·0001 μF. capacitance, and grid leak R_4, 50 000 ohms. If there is a tendency for the valve to set up a howl or low-pitched note, the value of resistance used as grid leak should be reduced, as this frequently eliminates the difficulty.

The screen grids are fed by a single resistance connected to the h.t. line, and decoupled by C_3 (0·1 μF.). R_2 is inserted in the position shown to make the stage more stable and prevent the generation of parasitic frequencies. Its value should be only a few ohms and not more than 5 ohms.

There should be an ample supply of anode current from the h.t. supply when the valve (A.C.TH.1) just described is to be used. When 250 anode volts and 100 screen-grid volts are applied to the hexode portion the emission current passing is 3·8 milliamps and 7·5 milliamps respectively, with 2½ volts on the control grid. To these figures must be added the anode current of the triode portion, which at 100 anode volts is 4 milliamps. The total cathode emission current (anodes and screen-grid currents) is thus 15·3 milliamps. This figure compares with about 6 to 10 milliamps that are usually passed by a mains heptode.

CHAPTER VI

THE OUTPUT STAGE

THE output stage of a receiver is that stage supplying the loud-speaker with power. It is called the power stage to distinguish its working from that of a normal voltage amplifier such as is used in a high-frequency stage or a low-frequency stage. In the case of the power stage, this has to convert the voltage applied to its input circuit into as much power as possible. In the production of power, the most important thing is to provide as high a current value as possible, for the power supplying the loudspeaker is proportional to the square of the current. In the case of voltage amplification, it has already been seen that the object is to provide as high a voltage in the output as possible, regardless of the actual amount of current. It should be pointed out that power is required to operate the usual types of loudspeaker employed in modern broadcast receivers. There are types of loudspeakers that require voltage as distinct from power, but these types have not attained any practical significance up to the present.

There are a number of difficulties that have to be overcome in the conversion of voltage into power. Foremost among these is the generation of spurious frequencies bearing a definite relation to the fundamental frequencies that are passed through the output stage to be reproduced. For example, suppose a 1 000-cycle note is required to be converted by the output stage to operate the loudspeaker. It is frequently found that under practical operating conditions there are reproduced not only the original frequency of 1 000 cycles per second, but also parasitic frequencies of 2 000, 3 000, and so on. These additional frequencies are known as harmonics and are found to occur in all types of output stages unless special precautions are taken. It is not an object of this book to give more than the briefest outline of theory, but it can be stated that if high quality reproduction is required, it is essential in the output stage to comply with one of two conditions—

(1) The output valve must have a power handling capability much greater than the average amount of power required from it; or

(2) A medium power valve must be used in such a way that it is only handling a small fraction of its maximum rating.

The main reason for using the output valve in the manner indicated is that the actual power passing through it varies over

THE OUTPUT STAGE

very wide ranges during the reception of broadcast signals. For example, in an ordinary symphonic orchestral reproduction, the output power appearing in the anode circuit of the valve may easily fluctuate in the ratio of thousands to one. It is obvious, therefore, that if the output valve is rated at, say, 3 watts and that it is operated in such a way that at average signal strength it is providing 1½ watts of power, then during passages of fairly loud reproduction the valve will be badly overloaded. When this occurs, the harmonic distortion already mentioned becomes very prominent and the reproduction is very poor as a consequence. By using a valve, therefore, that has an ample power handling capability, the main cause of harmonic distortion is overcome.

In addition to using a valve of ample size for the output stage another important point for quality reproduction is to have an ample supply of h.t. voltage. It can easily be demonstrated that the actual power capabilities of a valve are increased very considerably as the anode, and in the case of the pentode, the screen-grid voltage is raised. For example, there are several pentode valves on the market which give an output with screen and anode voltage at 100 volts, of 0·3 watts. As these screen and anode voltages are raised up to the maximum recommended by the manufacturers, i.e. 250 volts, the power output can be increased to 3 watts, or ten times that given with only 100 volts supply. It is not pretended here that the valve is intended to operate at 100 volts only on screen and anode, but these figures are given to illustrate the point that it is essential, to get satisfactory power output, and consequently distortionless reproduction, to have a high anode and screen voltage.

The power referred to above is the alternating current power and not the anode dissipation. These two factors are likely to be confused unless it is clearly understood that the anode dissipation rating of a valve is the d.c. power at the anode, whereas the a.c. power is that in the actual signal output of the valve. The ratio of a.c. power to anode dissipation varies in the different types of valve and in the triode is usually about one-fifth, whereas in the pentode it is about one-third.

A common form of output circuit using a triode is given in Fig. 26. Here the input transformer T_1 has applied to its primary winding the voltages from the previous stage which may be an l.f. amplifier or the detector valve. The voltages in the primary of T_1 set up corresponding voltages in the secondary winding and these are applied to the grid of the output valve. This variation in voltage at the grid controls the anode current flowing through the primary winding of T_2 connected to the anode. The alternating currents produced in the primary of the output transformer T_2 thereby set up corresponding fluctuations in the secondary of T_2. T_2 is the loud-speaker transformer, and the secondary winding is joined to the speech coil of the loudspeaker.

The transformer T_2 has a step-down ratio. In the preliminary notes in this chapter it was mentioned that power was required to operate the loudspeaker. The larger the current provided to the speech coil of the loudspeaker the greater will be the electromagnetic effect between the speech coil and the magnetic flux from the speaker field-winding, and consequently the louder will be the reproduced sound. In order to obtain this large current it is necessary to step-down the voltage from primary to secondary.

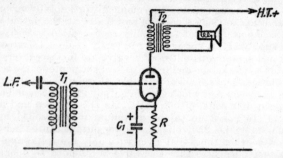

FIG. 26. THE TRIODE AS OUTPUT VALVE

The current in the secondary of the transformer is inversely proportional to the voltage for a given wattage. The actual value of the step-down ratio is dependent upon the valve and the speech coil of the loudspeaker. The effective primary impedance has to match the valve and the secondary impedance has to match the speech coil. The actual calculation of this ratio is, of course, a matter for the designer and is not relevant here.

The use of choke capacitance coupling for the loudspeaker is not a very popular one for the reason that it is more costly than the use of the circuits given elsewhere in this chapter. To be effective, the choke must have an inductance of not less than 20 henries when used in the circuit. This means that the anode current of the valve must be such that it does not diminish the inductance of the choke below 20 henries. This is important, for if the inductance is allowed to fall much below 20 henries, the effectiveness of the triode stage becomes seriously diminished.

One advantage of a choke-coupled circuit is that the loudspeaker transformer is isolated from the h.t. voltage. This advantage is of considerable importance when it is desired to use an external loudspeaker or any arrangement in which the wires that

THE OUTPUT STAGE

go into the loudspeaker transformer become exposed. In a choke-capacity circuit, it is perfectly safe to handle the loudspeaker, but in the circuits given in Fig. 26, this is not so, as there is at the loudspeaker transformer primary the full h.t. voltage of the receiver. An advantage of the choke capacitance coupling for l.f. amplifiers is that the utmost can be made of a limited supply of h.t. This is due to the low d.c. resistance of the choke, which in many cases amounts to only 300 or 400 ohms. The voltage drop down it is, therefore, very small and only a few volts from the h.t. supply are lost. When a low-frequency transformer is used, the primary winding usually has a resistance of somewhere in the region of 1 500 ohms. The d.c. voltage lost in this is, therefore, very much greater than that in the choke, and in cases where the anode voltage is not ample this is a serious drawback.

Still another advantage, but one which is not very important in modern broadcast receivers, is that by isolating the primary of T_2 from the anode current there is no disturbing effect on the core due to the anode current flow. It was mentioned in connection with low-frequency amplifiers that the flow of anode current in a transformer may have serious consequences. In the modern loudspeaker transformer, however, this drawback is not a serious one as the transformers are made sufficiently robust to withstand the current flow.

Grid biasing resistance R has a value which will depend upon the valve used and which can be calculated in accordance with the notes given in Chapter VIII. The condenser C_1 should have a large capacitance in order to by-pass the lowest frequencies in the signal without weakening them. Its capacitance may be up to 50 μF., and an electrolytic condenser may be used connected as shown with the positive terminal going to the cathode of the valve.

Pentode as Output Power Valve. The pentode is very popular as the output valve owing to its great sensitivity. By sensitivity is meant its ability to produce a large power output for a given voltage input. The reason that the pentode is so much more sensitive than the triode is that its amplification factor is so much greater. Many pentodes, for example, have an amplification factor of 150. This is many times greater than the amplification factor of a triode. When a large power output is required for a rather small signal input, the pentode becomes very useful. At the same time, however, it should be mentioned that distortion is likely to be much more noticeable with the pentode than with the triode. Nevertheless, if a pentode is operated well within its capabilities as a power reproducing valve, it can give an output which is not noticeably distorted. For example, if a 3-watt pentode is used and the average value of the power output is about

½ watt, the actual amount of distortion will generally be below that at which the average ear can detect it. Nevertheless, it must be admitted that when the greater efficiency of the pentode as compared with the triode is made use of, the distortion is greater than that of the triode.

To the average individual, however, the distortion occurring in a pentode power output stage is not very troublesome, especially

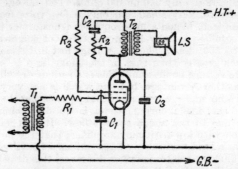

FIG. 27. PENTODE OUTPUT STAGE

if the precaution, already mentioned, of not allowing the valve to become overloaded is taken.

One circuit for the pentode is given in Fig. 27. An input transformer T_1 is shown, and in the grid lead is a resistance R_1. This resistance serves the purpose of stopping any h.f. currents from developing a voltage at the grid of the valve. R_1 does not weaken the desired audio signal if it is properly chosen. The actual value of R_1 is dependent upon the type of valve used. It will frequently be found that the value of R_1 can be 50 000 ohms or less. Up to 250 000 ohms may sometimes be used.

In the output circuit of the pentode is shown the primary of the output transformer T_2 connected directly between the h.t. supply and the anode. Across the primary is condenser C_2 and variable resistance R_2. C_2 and R_2 form a tone corrector and also serve to limit the load impedance of the valve at high audio frequencies, and thereby avoid the possibility of an excessively high voltage appearing at the pentode-anode.

Considering first the tone corrector C_2 R_2 it will be remembered that the harmonic distortion results in spurious frequencies higher than the fundamental appearing in the output circuit. Owing to the distortion produced by the pentode, these harmonic frequencies being higher than the original tones producing them,

THE OUTPUT STAGE

tend to make the music reproduced sound more high pitched. Unless some steps are taken, therefore, to diminish the intensity of the harmonics the distortion will be too noticeable. The tone corrector C_2 R_2 reduces the strength of the higher frequencies in the output by providing a shunt circuit to the primary winding of T_2 along which these higher frequencies may pass instead of through the primary winding itself. The value of C_2 may be 0·02 μF. and R_2 may be 20 000 ohms. As the potentiometer arm moves along R_2 towards the terminal connected to the condenser C_2, the effective capacitance in shunt to the transformer winding increases and correspondingly the proportion of the higher frequencies that are by-passed from the transformer primary. Music thereby becomes lower pitched, and this is precisely what is required. The amount of tone control is thus variable and the arrangement shown is used in practically all pentode circuits.

Additional by-passing can be obtained by having a capacitance across the valve shown at C_3 and in conjunction with the tone corrector outlined C_3 may have a capacitance of about 0·01 μF. capacitance.

The screen grid is shown connected to the h.t. through a resistance R_3. In practice it will frequently be found that R_3 is not necessary and that all that is required is to connect the screen directly to the h.t. supply. Sometimes, however, it is advisable to use R_3 in order to drop the h.t. down to that required for the screen. Even if the screen voltage required for the valve is the same as that of the anode, it should be remembered that there is a voltage drop down the primary of T_2 and unless R_3 is used the voltage actually at the screen will be higher than that at the anode. In most cases this is not of serious consequence, but the inclusion of R_3 in the circuit may sometimes be advisable. This resistance can be 2 000 to 5 000 ohms, depending upon the valve. The screen-grid decoupling condenser C_1 is also frequently unnecessary but may improve the operation of the valve in some circuits. The value of C_1 may be 1 or 2 μF.

An alternative circuit for use with the output pentode is given in Fig. 28. This shows the input to the pentode being fed directly from a diode. Usually, when a pentode is used in this way, a double-diode-pentode is used, but this is not necessary.

The main point of interest in this circuit is the tone control connected across the grid and cathode electrodes. Switch S connects one of three condensers, C_2, C_3, and C_4, across the grid leak R_3. The capacitance of C_2 is 0·0005 μF.; C_3, 0·0015 μF.; and C_4, 0·005 μF. As the capacitance across the grid leak is increased, the reproduction becomes lower in tone. It is not of great consequence whether the tone control is in the input or the output circuit but the arrangement shown in Fig. 28 is quite good. The grid leak resistance R_3, which may be ½ megohm in value, is tapped for use as a manual volume control.

The use of a tone control in the input circuit does not avoid the necessity for using a condenser across the output circuit. A condenser in the latter position is necessary in order to avoid an excessively high load impedance at the higher musical frequencies. The impedance of the primary of T increases with frequency and at, say, a frequency of 5 000 cycles, the impedance of the primary may be so great that an excessive a.c. voltage may appear at the

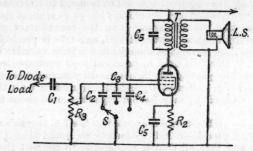

Fig. 28. Alternative Circuit for a Pentode Output Stage, showing Tone Control on Input Side

anode of the pentode valve. In order to avoid this, it is necessary to shunt the primary by condenser C_5, which is so dimensioned that whilst it will not cut off too great a proportion of the higher frequencies, it will prevent the effective output impedance becoming too great. A suitable value for C_5 is 0·001 µF., or even slightly higher.

It is noticed that the secondary of the output transformer T is connected to earth. This is a useful connection and helps to maintain the general stability of the circuit. A commonly used connection is to take a lead from the cathode end of the biasing resistance R_2 to the cathode of the a.v.c. diode. This applies to the cathode of the a.v.c. valve a positive voltage which is used for delaying the operation of the a.v.c. system. The use of this arrangement avoids the necessity for a special delay resistance.

In Fig. 29 is shown a further method of connecting an output pentode. In the input circuit is connected a high-frequency choke Ch and condenser C_1 in series across the grid leak. Ch and C_1 can be designed to act as a filter for disturbing heterodyne frequencies. The actual construction of Ch will be a matter for experiment and will depend upon the particular

THE OUTPUT STAGE

frequencies it is desired to eliminate. The most troublesome heterodyne frequencies are usually from 6 000 to 10 000 cycles per second and if a value of, say, 0·002 µF. is used for C_1, various sizes of choke Ch can be constructed to provide a suitable filter. R_1 is the usual grid-leak resistance and R_2 is an h.f. stopper resistance. This resistance R_2 is similar to R_1 given in Fig. 27.

The tone control in this circuit consists of condensers C_4, C_5, and C_6, which are connected across the valve by means of switch S. Suitable values for these condensers are C_4, 0·001 µF.; C_5,

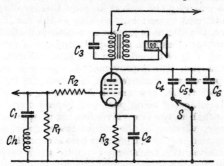

Fig. 29. Showing another Circuit for the Output Pentode

0·005 µF.; and C_6, 0·02 µF. When tone control condensers are used in this position they must be of the high insulation type owing to their having the full h.t. voltage applied to the pentode valve directly across them. This is one disadvantage to the tone control in the output circuit as compared with a tone control in the input circuit shown in Fig. 28.

The condenser across the primary of T in Fig. 29 is not so necessary in this case as in Fig. 28, as there is across the valve a tone control condenser which limits the impedance in the anode circuit of the pentode. On the other hand, it is as well to have a condenser in the position shown at C_3 in order that no harm may arise from a disconnection of the tone control condensers at any time or through a fault. The value of C_3 may be 0·0005 µF.

CHAPTER VII

PUSH-PULL AMPLIFIERS

A PUSH-PULL amplifier is one in which two valves are employed, the input voltages to these valves being in phase opposition but equal in amplitude, the signal output voltages being additive. The process whereby the valves amplify these voltages may be of several different types; but, so long as the conditions are as stated above, the circuit is known as a push pull amplifier.

The advantage of using a push-pull amplifier is primarily that a higher fidelity is obtained than if a straightforward amplifier circuit is employed. The improvement in quality is brought about by the output circuit arrangement in which the most seriously distorting factor, i.e. the second harmonics, is cancelled out. Another advantage of using push-pull is that owing to the d.c. anode currents from the two valves flowing in opposite directions in the primary winding of the output transformer, the effect of the anode current on the transformer core is cancelled out. It was seen in connection with transformer-coupled single valve amplifiers (see page 29) that the anode current flow in a transformer primary had a disturbing effect on the operation of the transformer. As against these advantages must be set the fact that with push-pull only half the total input voltage is applied to each valve. Although, therefore, each valve may handle its full grid swing, this is actually only half the total voltage provided by the previous amplifier. However, push-pull amplifiers find a very popular demand on account of the much higher fidelity that they provide.

Class A Push-pull Amplifier. A Class A push-pull amplifier is one in which the bias given to the valves is such that each is operated about the centre point of the linear portion of its characteristic. In practice, this arrangement is seldom faithfully complied with, as it is found that a higher efficiency can be obtained by using a slightly higher bias than a true Class A would require. By doing this, the grid swing available for each valve is correspondingly increased and the anode current is reduced. The distortion, which in a straightforward amplifier would be set up by operating the valves in this way, is cancelled out in the output circuit of the push-pull amplifier.

The circuit of a Class A push-pull amplifier is given in Fig. 30. The input voltage is applied across the secondary winding of the input transformer T_1 in the manner already described in connection with transformer-coupled amplifiers. It will be noticed

PUSH-PULL AMPLIFIERS

that the secondary winding S in this case is centre tapped and connected to earth. This in effect cuts the secondary winding in half, one half being connected to the input of the upper valve and the other half to the input of the lower valve. As the voltage across S fluctuates correspondingly to the current passing through the primary P, the voltage at the two ends of S varies, therefore, in the opposite sense. As the terminal of S connected to the upper valve becomes, say, 10 volts positive, so the terminal of S connected to the lower valve becomes 10 volts negative with respect to the earth line. This is, of course, the condition for push-pull operation as already defined.

In the output circuit the result of this voltage variation in opposite phase can be readily understood by noting the direction

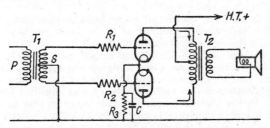

FIG. 30. A CLASS A PUSH-PULL OUTPUT STAGE

of the direct current flow from the anode of each valve as indicated by the arrows. The current from one valve passes through the top half of the primary of T_2 and downwards to the h.t. tapping, and the current from the other valve passes through the bottom half of the winding to the h.t. tapping. The effects of the two d.c. currents oppose each other on the core of the transformer and the harmful effects of the d.c. polarizing current are annulled. On the other hand, the alternating voltages induced across the primary are additive, so that the total a.c. voltage due to the signal is double that provided by one valve.

The step-up ratio of T_1 is usually fairly high in order to overcome the disadvantage already mentioned of applying only half of the total voltage in the secondary winding to each valve. The step-down ratio of the output transformer T_2 is governed as in all other cases where a loudspeaker is connected, by the matching impedance required for maximum output to the loudspeaker speech coil. This matter is considered in greater detail in the chapter on output circuits.

In practice it is usually an advantage to connect a resistance in series with each grid of the Class A push-pull amplifier as shown at

R_1 and R_2. These resistances make the amplifier a lot more stable than it would be without them and their use does not entail any disadvantage as regards a reduction in effective voltage at the grid. Their value is about 30 000 ohms.

Class B Push-pull. For the purposes of this book a Class B amplifier may be understood to be one in which the valve grids are so biased that each valve works on the bottom bend of its characteristic and in which the grids are driven positive during the operation of the amplifier. Theoretically, Class B push-pull amplifier includes any push-pull circuit in which each valve is worked at the bottom bend of its characteristic. This would include quiescent push-pull amplifiers described later on in this chapter. However, in the broadcast receiver industry, the Class B push-pull is understood to be a positively driven push-pull circuit of the type mentioned above and the quiescent push-pull amplifier is generally understood to be that given under the heading "Quiescent Push-pull Amplifier."

In considering Class A push-pull amplifiers, it was stated that when the valve grids are given a higher negative bias than that required to work them on the middle point of the linear portion of their characteristics, a higher efficiency was obtained from each valve. In the Class B amplifier this increase in bias in order to obtain a higher efficiency is carried to its logical conclusion and the bias is increased to such an extent that each valve operates on only one half of the cycle, leaving the other valve to operate on the alternate half-wave of the signal voltage. Furthermore, the Class B amplifier valve is especially designed so that the input signal can be allowed to drive the grids positive. In all ordinary amplifiers this would be fatal to the waveform in the output of the amplifier. By special construction of the valves and circuit, however, the harmful effect of positive grid potential and consequent grid current flow is reduced and there is obtained from the Class B valve an efficiency of somewhere in the region of 65 per cent, as compared with 23 per cent from a single amplifier valve operated on the mid-point of its characteristic.

The main uses of the Class B valve are in connection with battery receivers. By operating the valve well down its characteristic, the steady anode current flow that takes place whether a signal is being received or not, is greatly diminished. In fact with many practical Class B circuits the steady current flow is of the order of 1 milliampere as against the normal current flow when working the valve as a Class A amplifier of ten or more milliamperes. It follows, therefore, that when no signal is being applied to the amplifier, the consumption of the h.t. battery is very small. Furthermore, the actual consumption of h.t. current is roughly proportional to the signal intensity. If it is desired,

PUSH-PULL AMPLIFIERS

therefore, to economize the use of the h.t. batteries, all that it is required to do is to use a volume control at a stage previous to the Class B amplifier and to adjust it in such a way that the signal voltage applied to this amplifier is low. The high efficiency of the valves when operating a Class B circuit enables the battery user to obtain a receiver which will provide a comparatively large output without going to a big expense in the matter of large valves and correspondingly expensive h.t. source.

A simple but practical circuit is shown in Fig. 31. A Class B amplifier consists of the triode valves V_2 and V_3 and the input and output transformers. Considering this part of the circuit, it is

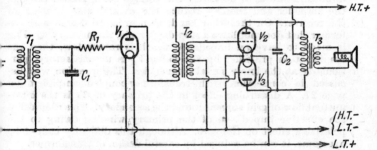

FIG. 31. CIRCUIT OF THE CLASS B OUTPUT STAGE, WITH DRIVER VALVE V_1

noticed that the centre point of the secondary of the input transformer T_2 is connected directly to the earth line. This means, of course, that there is no grid bias applied to the Class B valves V_2 and V_3. It should be mentioned here, however, that not all Class B valves are operated in this way. Several types of Class B valve require a grid bias voltage in order to operate on the lower bend of their characteristic. Whether a bias voltage is required or not depends entirely upon the construction of the valve itself. The signal voltage is applied to the control grids in opposite phase and equal in amplitude by the secondary of T_2 in a manner similar to that already outlined in connection with Fig. 30. Also, the outputs of V_2 and V_3 in the primary of the transformer T_3 produce similar effects to those noticed in considering the operation of Class A amplifier shown in Fig. 30.

The difference in working of the Class A and Class B amplifier is entirely in the situation of the operating point of the valves and in the input circuit. For example, when the grid is driven positive, a grid current will flow. In the case of a strong signal, the grid current may rise to as much as 25 milliamperes. If the resistance of the secondary of the transformer T_2 is fairly high, then a considerable voltage drop and consequently a power loss will be produced. It is essential, of course, that this should be avoided. The transformer T_2 is, consequently, a step-down transformer in which the primary winding is made to have a large impedance and the secondary winding as low a d.c. resistance as practicable. The d.c. resistance of the secondary winding of T_2 is usually from 200 to 450 ohms and the resistance of either half of the transformer will, of course, be half of this figure. In this way the power loss is reduced to a minimum. It is desirable, of course, that the step-down ratio should be kept low so as not to diminish more than necessary the voltage being passed through the amplifier and thereby reduce the overall amplification of the receiver. On the other hand, if in order to design a transformer that does not have a step-down ratio, the primary winding is made to have an impedance which is rather low, then the amplification provided by the valve V_1 is unsatisfactory, particularly at the lower signal frequencies. The question was discussed in connection with transformer-coupled amplifiers on page 27. A disturbing effect in the primary of T_2 is the intermittent flow of grid current round the secondary. This diminishes the effective impedance of the primary winding owing to the polarizing effect on the core. Although this difficulty cannot be overcome, it can be reduced by careful design of transformer.

The use of the valve V_1 is rendered necessary by the loss due to the positive driving of the grids of the Class B valves. In practice it provides the extra amplification that is usually required by a push-pull valve if this amplifier is to be operated over its entire grid swing. It was noted in connection with the Class A amplifier that the voltage of the secondary of the input transformer is divided between the two push-pull amplifier valves. As triodes are used in the Class B circuit and these triodes are not so sensitive as pentodes, it is advisable if the utmost use is to be made of the Class B amplifier to use the driver valve V_1. In Fig. 31 the input circuit to V_1 is shown as the secondary of an inter-valve transformer. It need not necessarily be coupled via a transformer to the detector stage, but may be coupled by resistance-capacitance.

The operation of Class B involves a certain distortion owing to the fact that it is extremely difficult to get a combined characteristic that is linear. Owing to this non-linearity in the combined characteristic, there frequently results a harshness in the reproduction commonly known as Class B shriek. In order to diminish this, a condenser is connected across the output transformer primary. In Fig. 31 this is shown at C_2 and has a capacitance of about 0·005 μF. There are a number of modifications that are possible to the circuit shown in Fig. 31. In fact, the circuit operation is frequently improved by the addition of a certain number of condensers as shown in Fig. 32.

Across the secondary of input transformer T_1 is connected a condenser value 0·0005 μF. for the purpose of stabilization, but is not necessary if adequate stability is provided in the anode circuit. Across the push-pull valves themselves are connected two condensers C_2 and C_3, each with a capacitance of 0·002 μF. for the purpose of preventing parasitic oscillation. C_4 also has a capacitance of 0·002 μF. These three condensers operate in association with C_5 to provide the necessary tone correction for the amplifier. The condenser C_5 has a capacitance of 0·005 μF., and in conjunction with the resistance R_2 of

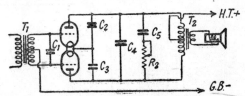

FIG. 32. AN ALTERNATIVE ARRANGEMENT FOR THE CLASS B AMPLIFIER

30 000 ohms resistance provides a manual control of the tone. As the amount of this resistance in the circuit is increased the tone of the reproduction becomes higher, and, conversely, when the resistance in circuit is reduced the tone becomes lower pitched.

In the operation of the Class B amplifier it should be remembered that if a grid bias is required by the valve it will be necessary to reduce the voltage applied to the grids when the h.t. battery voltage falls. If this is not done, serious distortion is noticeable in the reproduction.

It should not be thought that because the static current of a Class B amplifier is small, only a small capacity of h.t. battery will be required. Actually the peak value of the current consumption by the Class B valves may be very considerable; for example, up to 40 milliamperes. If, therefore, a low capacity battery is used, a severe drain would be put upon it and its useful life would be short. Alternatively, the Class B amplifier will not be operated satisfactorily and serious distortion will be experienced after the small capacity battery has been in service for a short time. If the anode current flowing in the Class B amplifier is checked up during the operation of the receiver, the reading will naturally fluctuate in accordance with the intensity of the signal being amplified. It follows, therefore, that the usual indication of distortion in a

Class A amplifier, i.e. a fluctuating anode current, will not be effective in this case.

Quiescent Push-pull Amplifier. A quiescent push-pull amplifier is an amplifier operating in push-pull, and in which a grid bias is used of such a value that the valves are operated on the bottom bend of their characteristics and in which the grids are never allowed to become positive. The operating point on the characteristic is in a similar position to that employed by the Class B amplifier, but the grids are never driven positive.

Pentode valves are usually employed as quiescent push-pull amplifiers, the two sets of valve electrodes constituting one

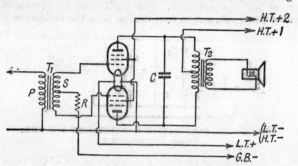

Fig. 33. The Quiescent Push-pull (Q.P.P.) Output Circuit

quiescent push-pull valve. Owing to the high slope of the pentode valves, a driving valve is usually unnecessary in order to operate over the full grid swing of the valves. Furthermore, as the grids in this case never become positive, there is no loss of power owing to the flow of grid current through the input transformer secondary winding. As compared with the Class B amplifier the input circuit is greatly simplified owing to this lack of grid current. The input transformer is very similar in design to that of a Class A push-pull amplifier and usually has a step-up ratio. In fact a high step-up is usually employed and this represents a considerable advantage as compared with the Class B circuit. This is one of the reasons that has resulted in quiescent push-pull amplification being much more commonly used than a Class B push-pull amplifier.

The circuit arrangement is given in Fig. 33. The primary of the input transformer T_1 is joined to the detector valve in the usual manner, or it may be conveniently connected to the triode of a double-diode-triode valve. The stepped-up voltages in the secondary winding of T_1 are applied to the control grids in the

PUSH-PULL AMPLIFIERS

usual manner for push-pull operation and the amplified voltages appearing in the output circuit are applied in additive relation to the primary of the output transformer T_2. The centre tapping of the secondary winding S on T_1 is taken via a resistance R to a suitable negative bias voltage.

R is in the common grid return of the valves and stabilizes the circuit by preventing any tendency for the amplifier to break into oscillation. The value of R varies with the different types of valve, from 100 000 ohms to 250 000 ohms. The value in any event is not very critical, but should the stage be inclined to be

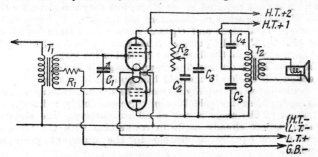

Fig. 34. An Alternative Q.P.P. Stage

unstable, the value of R employed should be increased. The distortion already explained in connection with the Class B amplifier brought about by inexactitudes in the matching of the two characteristic curves is also present with the quiescent push-pull amplifier.

When pentodes are used, harmonic distortion is more prevalent than when triodes are used. It was mentioned earlier in this chapter that the even order harmonics were cancelled out in the output circuit of a push-pull amplifier. It follows, therefore, that in the quiescent push-pull amplifier, although pentodes are used, the disturbing second and fourth harmonics will not have such harmful effects on the quality of the reproduced signal as in a straight amplifier. Unfortunately, however, the odd harmonics are produced by pentodes and these remain in undiminished intensity in the output of the quiescent push-pull amplifier. This results in the reproduction being high pitched, and these odd order harmonics increase still further the harshness in the reproduction. It is imperative, therefore, that a shunt condenser of large enough capacitance be connected across the primary of the output transformer T_2. In Fig. 33 this is shown at C, and has a capacitance of 0·001 μF.

A number of refinements can be added to the circuit shown in Fig. 33, mostly modifications similar to those associated with the Class B amplifier. For a further explanation, a number of additions to the circuit already outlined are given in Fig. 34. Across the secondary of the input transformer T_1 is connected a variable condenser C_1. This acts as a variable tone control and may have a maximum capacitance of 0·00075 μF. This condenser is a little unusual, but if it can be fitted in the position shown it will be found to afford a useful control of the tone. An alternative tone control is shown at R_2 and C_2. It will be obvious that both tone controls are not required in one stage, but they are given for the sake of showing what modifications are feasible. R_2 has a value of about 50 000 ohms and C_2 may have a capacitance of about 0·01 μF. C_3, C_4, and C_5 may each have a capacitance of 0·001 μF. and then constitute the fixed tone control of the stage.

In operating a quiescent push-pull stage, it is very necessary to watch rather closely the grid bias applied to the valves. The harmonic distortion rapidly increases as the h.t. battery voltage falls and, unless the grid bias is correspondingly reduced, the operation of the stage is not at all satisfactory. The remarks in this respect already made in connection with the Class B amplifier are equally applicable to the quiescent push-pull amplifier.

Paraphase Push-pull Amplification. The circuit termed a paraphase push-pull amplifier is one in which the opposition in phase necessary for the operation of the push-pull amplifier is obtained by means of a valve. It is a basic principle of the operation of thermionic valves that the a.c. voltage at the anode is in opposite phase to the a.c. voltage at the grid. If, therefore, some means is provided for tapping off the voltages at the grid and anode of a valve, these two voltages must be in opposite phase. Provided some means is available for rendering the amplitude of these two voltages equal, the resultant voltages will then be suitable for application to a push-pull amplifier. This is in effect what is done in the paraphase push-pull amplifier circuit.

The advantage of this type of circuit is the avoidance of the use of iron-cored transformers. For the highest quality in reproduction it is difficult to design a low-frequency transformer that will give a satisfactory transference of the voltage; there nearly always results distortion at one or both ends of the audio-frequency scale. By using a paraphase circuit, that is to say by replacing the input transformer by a thermionic valve, this disadvantage is overcome and instead of a transformer there is connected the valve and resistances. Much less distortion is brought about by these components, and the resultant overall frequency-amplification characteristic is very much more satisfactory.

PUSH-PULL AMPLIFIERS

One circuit for providing paraphase push-pull amplification is shown in Fig. 35. The low-frequency voltages, say, from the detector valve, are applied to the first valve in the usual manner, R_1 being a normal grid leak. In the resistance R_2 in the anode circuit of this valve, there appear amplified low-frequency voltages. Coupling condenser C_2 is connected to one of the output terminals, T_1, and the amplified low-frequency voltage is passed from there to the subsequent amplifier. A portion of the low-frequency voltage along R_2 is tapped off by the variable tapping shown and is passed through coupling condenser C_3 to the control

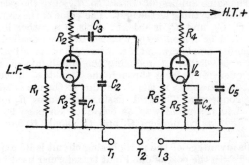

FIG. 35. CIRCUIT OF A PARAPHASING STAGE FOR SUPPLYING A PUSH-PULL AMPLIFIER

grid of the second amplifier. The point along R_2 from which these voltages are tapped must be such that the actual signal voltages of the grids of the two valves are exactly of the same amplitude. Now these voltages at the grid of V_2 are of the same phase as those passed through C_2 to T_1. However, at the anode of V_2, the voltages are opposite in phase to those at the input grid of this valve and, as a consequence, the voltage passed through coupling condenser C_5 to the third output terminal T_3 is in opposite phase not only to that at the grid of this valve, but also to that at the anode of the first valve and T_1. It is clear, therefore, that with respect to the earth connection joined to T_2 the signal voltages at T_1 and T_3 fluctuate in opposite phase and, owing to the tapping along R_2, these voltages are equal in amplitude. There are thus obtained suitable voltages for applying to a push-pull amplifier.

The push-pull amplifier associated with the paraphasing circuit is not shown in connection with Fig. 35 as it may be quite the normal one. This circuit is in fact merely the equivalent of the transformer T_1 of Fig. 30. Class A push-pull amplifiers are used with the paraphasing arrangement as the latter is essentially a high quality circuit. The Class A push-pull amplifier is the arrangement that provides the highest quality reproduction, and if a Class B amplifier, for example, were connected to a paraphasing circuit the main advantage of using the paraphase arrangement would be nullified by the reduced fidelity brought about by the Class B circuit.

The value of R_2 should be twice or three times the anode a.c. resistance of the valve. Suppose, for example, that the anode a.c. resistance is 10 000 ohms. R_2 can then satisfactorily be 30 000 ohms. The same applies, of course, to R_4, since the second valve is doing the same thing as the first. The value of the coupling condensers C_2 C_3, and C_5 is 0·1 μF. This is a rather high value, but it should be remembered that we are now dealing with the higher quality amplifiers and if a lower value than 0·1 μF. is used, there is a possibility of reducing the amplitude of the lower audio-frequencies passing through the amplifier. There is very little risk of instability in the circuit due to the use of such a high coupling condenser. The grid biasing resistances R_3 and R_5 are calculated in the usual way as described on page 83, and the value of by-pass condensers C_1 and C_4 should be not less than 6 μF.

The disadvantage of the paraphasing circuit is its expense. As compared with the cost of one input transformer used in the normal Class A circuit, there are in the paraphase circuit two valves five condensers, and six resistances. It should be noted that the arrangement shown in Fig. 35 is definitely not a two-valve amplifier circuit, for, owing to the necessary adjustment of the tapping along R_2, the voltage applied to the second valve is exactly the same as that applied to the first. The circuit gives the gain, therefore, of one resistance capacitance-coupled stage only. Another disadvantage is that the two valves have to be matched very carefully. If, for example, the emission of one valve deteriorates slightly more than the other, then the voltages provided from the two anode circuits will not be of precisely the same amplitude. The phases will still be in opposition, but owing to the reduced amplitude of one half-wave, the following push-pull amplifier is not fed with the correct voltage and distortion becomes apparent.

Single Phase Splitter Stage. The disadvantage last mentioned can be overcome to a certain extent by the use of the circuit arrangement shown in Fig. 36. In this arrangement, instead of there being used two valves for obtaining the phase opposition in voltage, the load resistance of one valve is split into two halves and the consequent opposition in phase at the cathode and anode is applied to the push-pull stage. The valve is known as a phase

PUSH-PULL AMPLIFIERS

splitter or a phase inverter. Referring to Fig. 36, it is seen that in the anode circuit is R_2 and in the cathode connection to earth are resistances R_3 and R_4. Now these resistances should be so dimensioned that R_3 and R_4 together are equal to R_2. Under these conditions the fluctuating anode current will produce voltage drops in the respective anode and cathode connections which are equal in amplitude.

Making connections with the anode and cathode of the valve ensures that the voltages applied to the push-pull amplifier are also opposite in phase. That this is so can also be seen from the following consideration. Assume, for example, that the signal voltage applied to V_1 becomes positive. The anode current flowing in the circuit R_2, h.t. source, R_3 R_4 to cathode, becomes

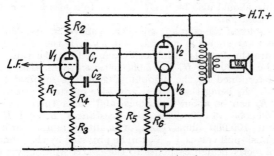

FIG. 36. A PUSH-PULL AMPLIFIER WITH PHASE SPLITTER VALVE

larger. The voltage drop along R_2, therefore, is increased and the anode becomes less positive with respect to earth. On the other hand, the voltage drops in the cathode resistances R_3 and R_4, being correspondingly increased, result in the cathode being less negative with respect to the earth wire. In other words, as the anode a.c. voltage relative to earth becomes negative, so the cathode becomes more positive. When a negative half-wave of voltage is applied to the control grid of V_1, the reverse operation takes place, that is to say, the anode becomes more positive with respect to the earth and the cathode becomes more negative.

It will be obvious from examination of Fig. 36 that the main disadvantage mentioned in connection with Fig. 35, i.e. regarding the possible dissimilarity in the emission of the two valves, is not present. For provision of equal inputs to the push-pull valves resistances R_3 and R_4 together should be equal to R_2 and this is not usually particularly difficult to arrange. There is still no amplification from the phase-splitter valve. In fact with this circuit there is a slight loss in amplification owing to a reverse reaction effect. This reverse reaction, usually termed degeneration, is brought about by the connection of the signal voltage input between the grid of V_1 and earth, as distinct from the more usual arrangement of input circuit between grid and cathode. This degeneration is due to the resistances R_3 and R_4 being common to both grid and anode circuits. Voltages in these circuits, being in opposite phase, tend to cancel out in the common resistances. One consequence of this degeneration is that the voltage applied to the phase splitter valve must be of sufficient magnitude fully to load the push-pull valves V_2 and V_3. It is necessary, therefore, in order to make the utmost use of this push-pull circuit to have an audio-frequency amplifier in order to obtain the requisite voltage. Allowance should also be made for a certain loss in gain due to the degeneration.

The values of the coupling condensers shown in Fig. 36 are the same as those shown in Fig. 35, i.e. 0·1 μF. If the phase splitter valves are the same type in both circuits, then the value of R_2 should be 15 000 and R_3 R_4 15 000 between them. The value of R_4 is determined by the amount of grid bias required for the valve V_1, R_4 being the normal grid bias resistance. In a usual circuit R_4 can be about 500 ohms, and under these conditions R_3 should be 14 500. The grid resistances R_5 and R_6 may be about 100 000 ohms each or a little more. Grid bias for the push-pull valves V_2 V_3 can be provided by the connection of the directly heated cathodes to the heater winding via a bias resistance.

CHAPTER VIII

POWER SUPPLY CIRCUITS

INCLUDED in this chapter are not only circuits for providing h.t. supply, but also supplies for heater and grid bias. Several of the circuits described in this chapter are used in the circuits outlined earlier in the book. For example, automatic grid bias or the method of obtaining grid bias by means of a resistance connected in the cathode lead of the valve is explained. In most of the circuits in this book that are adapted to be supplied by the electric mains, automatic grid bias is employed, but the present chapter is considered to be the relevant one for explaining the process whereby this is obtained.

Valve rectifiers are employed very considerably in broadcast receivers, primarily owing to their cheapness. For a given output in current and voltage they are cheaper than a metal rectifier and are much easier to replace when their useful life is finished. Another advantage of the use of valve rectifiers is that they take up much less room than a metal rectifier capable of producing the equivalent power. On the other hand, however, the metal rectifier is more robust and is not broken so easily as the fragile thermionic valve with a glass envelope. The life of the metal rectifier is very much greater than that of the thermionic valve, and the metal rectifier will also stand more overload than will its thermionic counterpart.

Power Supply from Alternating Current Mains

USE OF VALVE RECTIFIER. A simple but nevertheless very practical circuit for using the valve rectifier to supply h.t. and heater current is given in Fig. 37. The alternating current input from the mains is applied to the primary of the mains transformer T, this circuit being controlled by the make-and-break switch S. The alternating current flowing through P sets up a fluctuating magnetic field which cuts the secondary windings and in doing so produces a current of similar waveform in these three windings. The current thus produced in the secondaries S_1, S_2, and S_3 is, therefore, alternating. The main secondary winding S_1 has its respective ends connected to the two anodes of a double-diode rectifier valve V_r and the centre of S_1 is connected to earth. The two anodes of V_r will, therefore, become alternately positive as the voltage at the ends of S_1 fluctuates with the alternations of magnetic flux induced into it by P. The electron emission from the cathode of V_r is attracted to whichever rectifier anode is positive and only while such anode is positive. Consequently the electron current from rectifier cathode passes to the anodes and then through the respective halves of the secondary winding S_1 to earth and back to rectifier cathode through the receiver valves and h.t. + line.

Since the electrons, which are negative in sign, pass from cathode to anode inside the valve, it follows that the cathode must, during emission, be at a positive d.c. potential with respect to the anodes of V_r. It may seem perhaps rather strange that notwithstanding

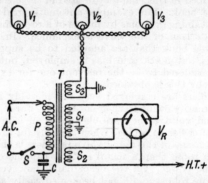

Fig. 37. A Mains Supply Circuit for A.C. Voltage

the high a.c. potential from the secondary S_1 is applied to the anodes of V_r, the cathode is the positive d.c. electrode. Since, however, as just mentioned, the electrons are negative and, therefore, the cathode is giving up its negative elements, the cathode itself must be positive so far as the external d.c. circuit is concerned. The h.t. voltage for the supply to the receiver valve is, therefore, taken from the cathode of the valve rectifier as indicated in Fig. 37.

Additional secondary windings are shown at S_2 and S_3. S_2 supplies the alternating current to the valve rectifier and S_3 supplies the alternating current to the heaters of the receiver valves which in Fig. 37 are represented at V_1, V_2, and V_3. Unless some steps are taken to earth the winding S_3, it will be found that considerable hum will develop in the receiver. Many transformers have this secondary S_3 centre tapped and this centre tap must, of course, be connected to earth. Sometimes it happens that this tapping is not connected in the true electrical centre. To overcome this difficulty, a low resistance potentiometer should be connected across the two terminals of the secondary winding

POWER SUPPLY CIRCUITS 77

S_3 and adjusted as explained in connection with Fig. 38. When a potentiometer is connected across S_3, the centre tapping on S_3 should be disconnected from earth.

It is seen that the heaters of the receiver valves are connected in parallel. A common rating of a.c. heaters is 1 ampere per valve. In a four-valve receiver there will thus be 4 amperes of current passing from S_3 to the receiver heaters. If, therefore, there is any resistance in this part a serious drop of voltage (equal to current times resistance) is likely to take place. When wiring this part of the circuit it is, therefore, necessary to use wire of high current-

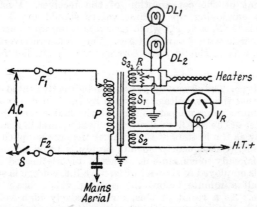

FIG. 38. AN ALTERNATIVE A.C. MAINS SUPPLY CIRCUIT

carrying capacity and to ensure that all the connections are well soldered or well joined. The object of using twisted wire from S_3 to the heaters is to annul the electro-magnetic field set up by the alternating current flowing through the wires. If this is not done, there is a likelihood of interference being set up at the frequency of the current passing through the winding S_3. As most a.c. mains have a frequency of 50 cycles, the frequency of the hum will also be 50 cycles and this will be reproduced by the receiver.

It sometimes happens that a certain amount of high-frequency current is induced into the secondary windings of the mains transformer from the mains. A simple method of overcoming this is to join a small condenser between the end of the primary of the mains transformer and earth. A condenser for this purpose is shown in Fig. 37 at C, this being of about 0·01 μF. The insulation of this condenser must, of course, be well above that of the mains supply so that risk of its breaking down is a minimum.

An alternative circuit for the mains unit is given in Fig. 38. This circuit gives the same result as the arrangement shown in Fig. 37, but is illustrated as an alternative circuit for providing high tension and heater current. In the circuit of the primary winding of the mains transformer are connected two fuses, F_1 and F_2. It is highly desirable that fuses be used in a mains unit for, owing to the high voltages that are present, there is quite a possibility of the entire receiver being set on fire by a short circuit. The use of fuses reduces this possibility and fuses should be fitted on all mains equipments. The actual rating should be several times that of the consumption of the receiver. Most valve receivers employing up to four valves should use $\frac{1}{4}$ ampere fuses for both F_1 and F_2, while receivers using up to six valves should be safe with $\frac{1}{2}$ ampere fuses. Very small receivers, using say two valves, should have correspondingly smaller fuses, of course.

The valve rectifier shown in Fig. 38 is of the indirectly heated cathode type. The advantage of using an indirectly heated cathode valve is that the h.t. voltage provided by the unit does not reach its maximum value almost as soon as the mains are connected. The heater of the rectifier valve has to heat up first in exactly the same way as the heaters of the receiver that are supplied by the power unit. Consequently no h.t. supply is provided until the receiver is ready to consume it. When a directly heated cathode rectifier is employed as shown in Fig. 37, a h.t. voltage is supplied about half a minute before the receiver valves have become operative. As a result of this, an excessively high voltage is provided in the circuit and this increases the risk of a breakdown in the condensers or insulation. The use of the indirectly heated valve shown in Fig. 38 overcomes this risk.

Across the secondary winding S_3 that supplies the valve heaters is shown a variable potentiometer R. This is for the purpose of providing a centre tap to the secondary winding S_3 for connection to earth and is in place of the earthed connection shown at S_3 in Fig. 37. The resistance of R need not be very great, say somewhere in the region of 25 to 30 ohms, and the resistance wire need not be particularly thick. With only 4 volts across S_3, 30 ohms resistance across S_3 will only pass $4/30 = 0\cdot 13$ of an ampere. In practice the position of the tapping along R is adjusted when the receiver is connected. As the position of the contact arm moves from one end of the resistance to the other, the hum reproduced in the receiver will be found to diminish to a minimum point when the arm is at centre, and increase again as the arm is moved farther round. The correct operating point is, of course, the position of minimum hum.

Across the valve heater supply wires are shown two dial lamps

POWER SUPPLY CIRCUITS

DL_1 and DL_2. These are connected in parallel, but if suitable lamps are obtained they may be connected in series with each other. When using dial lamps, it should be remembered that they consume a small amount of current. In most cases this will be very small as compared with that of a valve and may be neglected. Some lamps, however, consume a current which may not always be neglected and if three or, as in some cases, more dial lamps are used, their consumption should not be disregarded. It should also be noted that as the dial lamps are across the heater supply circuit, any intermittent connections in the lamp sockets, etc., will cause interference with the heater current and thereby set

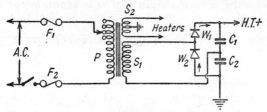

FIG. 39. CIRCUIT FOR USING THE METAL RECTIFIER IN THE MAINS SUPPLY UNIT

up noises in the receiver. For trouble-free operation of the equipment shown, therefore, the dial lamps should be well screwed in their sockets and any possibility of loose connections or short circuits should be avoided.

It is sometimes convenient to use an alternative aerial to that normally employed by the receiver, and the electric supply mains are useful in this respect. All that is required is a small condenser connected from one end of the mains transformer primary through a condenser to the aerial terminal on the receiver. A condenser for this purpose is seen in Fig. 38. This condenser should have a capacitance of $0{\cdot}0001$ μF. to $0{\cdot}0005$ μF. and should have a good insulation to avoid the risk of short-circuiting the mains supply to the unit.

Power Equipment Employing Metal Rectifier. Power units in which a metal rectifier is employed are simpler to construct than units employing the valve rectifier. This is because there is no heater circuit for the metal rectifier and no valve holders are required for connections.

The circuit of the supply equipment is given in Fig. 39. A transformer is employed for isolating the mains from the receiver and for stepping up the voltage. Although it is practicable from the point of view of supply of h.t. current to connect the mains directly on to the metal rectifier, this is a dangerous practice.

It is seen from Fig. 39 that the secondary winding S_1 of the mains transformer is connected into a bridge circuit consisting of two condensers C_1 and C_2 and two metal rectifiers W_1 and W_2. By convention the direction of the symbol for rectifiers of the type being considered is that in which positive current is allowed to pass. Tracing out the direction of the current, therefore, in the circuit containing the metal rectifier, it will be seen that across each condenser C_1 and C_2 there appears the rectified voltage provided by one rectifier. These voltages at the condensers are additive, and there is thus provided at the terminals of the h.t. supply a voltage double that obtained from one rectifier. For this reason the figure shown in Fig. 39 is known as the voltage

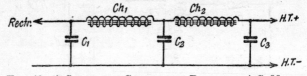

Fig. 40. A Smoothing Circuit for Rectified A.C. Mains Supply

doubler circuit. The condensers C_1 and C_2 actually form part of the circuit designed to smooth out the inequalities in the h.t. supply, and their action will be considered in connection with Fig. 40.

The secondary winding S_2 is a heater winding and operates in a similar manner to the windings shown in Figs. 37 and 38 as S_3.

Smoothing Circuits. The circuits considered up till now in this chapter have been the basic supply circuits for rectifying the alternating current received from the mains transformer. The d.c. voltage delivered by the rectifier is a series of pulses. If these were used for h.t., the receiver would function only during the periods of these voltage pulses. This arrangement is, of course, quite impracticable. The voltage required for operating a receiver must be substantially constant in amplitude. That is to say, the voltage provided by the mains unit must maintain a constant level, otherwise the irregularities in the h.t. supply will cause an interfering hum in the loudspeaker.

It is fortunately very simple to smooth out these pulses into a constant level of voltage. All that needs to be done is to connect a condenser of fairly large capacitance directly across the supply leads from the rectifier, i.e. between h.t. positive and earth in Figs. 37, 38, and 39. The current pulses provided by the rectifier then charge up this condenser, shown at C_1 in Fig. 40. This condenser at C_1 will begin to discharge through the receiver

POWER SUPPLY CIRCUITS

valves as soon as the voltage pulse from the rectifier falls below the voltage acquired by the condenser. However, the discharge of C_1 is so slow, if the capacitance of C_1 is suitably chosen, that before the charge in the condenser falls very appreciably another current pulse is supplied by the rectifier and the charge in C_1 is again raised to an extent determined by the current charging pulse and the load imposed by the receiver valves. This slight discharge and subsequent recharge by the following current pulse from the rectifier takes place so long as current is supplied by the rectifier. It will be apparent that if C_1 is too small, then the proportional discharge between the current pulses from the rectifier will be much greater than if a larger condenser is used. The condenser C_1 is called the reservoir condenser owing to its function as an electric reservoir.

The slight irregularities in the voltage level provided by the reservoir condenser have to be smoothed out for satisfactory operation of the radio receiver. It is impracticable to employ a condenser C_1 of such a size that the whole of the ripple is smoothed out, and consequently additional smoothing means have to be provided. The usual method of smoothing out the remaining ripples is to employ one or more chokes in conjunction with smoothing condensers. One circuit employing this is given in Fig. 40, where two power chokes Ch_1 and Ch_2 by-passed by smoothing condensers C_2 and C_3 are employed. The passage of the current through each choke sets up a magnetic field which tends to oppose the force producing it. When any variations in the current take place, therefore, such as during the slight ripple in the voltage from C_1, the magnetic fields of the chokes Ch_1 and Ch_2 tend to cancel out the alterations in current. The additional effect of the smoothing condensers C_2 and C_3 is sufficient to eliminate for all practical purposes the remaining ripple fluctuation in the h.t. voltage provided to the receiver. Most receivers employing four or five valves do not need to use two chokes and smoothing condensers as shown in Fig. 40. It is sufficient in these cases to use one choke and smoothing condenser of adequate dimensions.

The important point about the use of a choke for smoothing is that its inductance should be adequate at the current desired to be passed through it. The inductance of most chokes is inversely proportional to the flow of current through them. In selecting the smoothing choke, therefore, it is essential that the inductance at the rated h.t. current should be known. For the small type of receiver an inductance of 20 henries at the maximum current flow is found sufficient. With most chokes used in radio power units the inductance falls off rather rapidly as the current flow is increased above that at which the inductance is the rated 20 henries. Some chokes are specially constructed to give a fairly level inductance value over a large range of current flow; one common method of design for this purpose employs a core with an air gap in it.

The size of the smoothing condenser C_2 should not be less than 4 μF. and should preferably be more. In practice, values of capacitance up to 32 μF. are used for this purpose. If a receiver employs more than five valves, it is sometimes an advantage to use two chokes as shown in Fig. 40, and use an extra smoothing condenser at C_3. Furthermore, a connection to a certain number of the valves or say the output stage, can be taken from the junction of the two chokes, and the h.t. supply for the other valves from the terminals of the smoothing condenser C_3.

The loudspeaker field winding is commonly employed as a smoothing choke. The winding employed for providing the strong magnetic field required by the loudspeaker is of a sufficiently high inductance to make it a very satisfactory smoothing choke. The use of the loudspeaker field winding for this purpose results in a saving of the cost of a smoothing choke. It should, nevertheless, be borne in mind that the resistance of the loudspeaker field winding is usually a lot higher than that of a smoothing choke, and allowance will have to be made for the comparatively high voltage drop thereby brought about.

Sources of Grid Bias. There are various means whereby grid bias can be provided in a receiver other than by the simple employment of grid bias batteries. It is possible, for example, to use the anode current flow from the valves to provide a voltage drop down a resistance and to use this voltage drop for biasing the valves as required. This is practicable for both mains and battery receivers. Another method of obtaining grid bias is to use the normal voltage drop down a component in the receiver for purposes of biasing either directly or indirectly by means of a potentiometer connected across the component referred to.

Resistance in Cathode Lead. This arrangement can best be described by referring to Fig. 41. The arrows in that diagram indicate the flow of the electron current proceeding from the cathode. The electron current passes from cathode through the grids to the anode and from there through the output or the load circuit Ld, voltage dropping resistance R_2, h.t. supply source to the earth lead, back through the resistance R to cathode. The resistance used for the purpose of providing grid bias is R. It is seen that as the anode and screen grid current from the valve must pass through this to get back to the cathode, a voltage will be set up between the cathode and the earth line that is dependent on the resistance of R.

For example, suppose that 10 milliamperes of current is passing round the circuit from cathode to anode, external circuit back to cathode. If the value of R is 1 000 ohms, then the voltage drop (equivalent to the current times the resistance) must be 10/1 000 × 1 000 = 10 volts. If the resistance R is 2 500 ohms, then the voltage drop along it is 10/1 000 × 2 500 = 25 volts.

Now this electron current is proceeding from the cathode and, if it is remembered that the valve acts as a d.c. resistance by virtue of this emission, then it will be clear that since the cathode is between h.t. positive and h.t. negative, the earth line must be negative with respect to cathode by the amount of the voltage drop down the cathode resistance R. In the numerical examples given above the cathode will be in the first case 10 volts positive with respect to the earth line, and in the second case 25 volts positive with respect to the earth line. In order that the voltage drop along R can be used for grid bias, all that is necessary then is to connect the grid to earth;

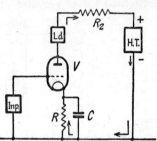

FIG. 41. ILLUSTRATING HOW AUTOMATIC GRID BIAS IS PROVIDED

and since the cathode is positive with respect to earth, it will automatically be positive with respect to the grid, which is, of course, the same as making the grid negative with respect to cathode. Negative grid bias is thus obtained and, as the process is automatic, this circuit arrangement is known as automatic grid bias.

It should be quite clear, however, that there is not the slightest justification for the title "free" grid bias in this arrangement, since the voltage in the cathode resistance is obtained at the expense of the voltage at the anode. For example, if the h.t. supply is 250 volts, then there are only 250 volts between h.t. positive and h.t. negative in Fig. 41. No matter how this voltage is distributed, it can never exceed 250 volts. If, therefore, 25 volts are dropped in resistance R, then only 225 volts are available for the rest of the circuit consisting of valve, load Ld and voltage dropping resistance R_2.

It will be apparent from Fig. 41 that R is common to both grid and anode circuits. As explained on page 74, this results in degeneration and loss of amplification, and so a means has to be provided to avoid this. Condenser C, if of sufficient capacity, will provide a low impedance path for the l.f. currents and thereby prevent degeneration by maintaining the cathode at earth potential as regards a.c.

In the case of high-frequency amplifiers the value of C need not be very large. A capacitance of 0·1 µF. to 0·25 µF. is quite satisfactory, but, as high-frequency currents are being dealt with, the by-pass condenser should be of the non-inductive type. With low-frequency amplifiers, however, a much larger capacitance will be necessary to be really effective. In fact, as the size of by-pass condenser C is increased from say 0·5 µF. up to some 25 µF. its effectiveness becomes greater. This effect is not a linear one, and

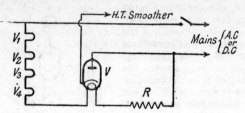

Fig. 42. Simple A.C./D.C. Mains Supply Circuit

after the capacitance is increased to about 12 µF., the influence of any further increase in capacitance is not very great. If a low value of capacitance is used for C, the lower frequencies in the audio range passing through the amplifier are not reproduced at full amplitude. Six microfarads capacitance should be the minimum value for C in low-frequency amplifiers. As the alternating component of the current in this circuit is very small, an electrolytic condenser may be used at C, condenser + being joined to cathode.

Power Supply in A.C./D.C. Receivers. When universal valves, or valves capable of operation from a.c. or d.c. mains are employed, a totally different type of power supply is necessary. Instead of the heaters being all connected in parallel as in the cases so far considered, they are connected in series. A rectifier is joined in series with one of the mains leads and this acts as an ohmic resistance when direct current mains are connected to the unit and as a rectifier valve when a.c. mains are connected to it.

The basic circuit of the a.c./d.c. mains unit is shown in Fig. 42. Considering the lower mains connection, this passes directly to the anode of the indirectly heated rectifier valve V and also through a resistance R, heater of V, and then through the heaters of the receiver valves shown at V_1 to V_4 and back to the other mains lead. The resistance R is known as the ballast resistance and is for the purpose of adapting the mains supply to that suitable for the valve heaters. For example, suppose the mains supply is 220 volts and that the valve heaters are rated at 13 volts. In the

POWER SUPPLY CIRCUITS

series circuit between the mains are the four heaters of the receiver valves and one of the rectifier valve, making five in all, each requiring a voltage drop of 13 volts. The total voltage drop down the heaters is thus $13 \times 5 = 65$ volts. The purpose of R is, therefore, to reduce the 220-volt supply to 65 volts so that an excessive current does not pass through the valve heaters.

The h.t. supply for the receiver is taken from the cathode of the rectifier valve V. When d.c. mains are connected to this unit the space between the cathode and the anode of the rectifier valve V is merely a resistance in the path of the current. The positive voltage path is, therefore, from positive main to anode, from anode to cathode inside the valve and cathode to h.t. supply. With a.c. supplied to the mains leads, the valve V acts as a half-wave rectifier. As the anode of V becomes positive at every alternate half-wave, so an electron current passes to the anode and makes the cathode positive, rectification taking place as a series of pulses at the same frequency as the mains supply. It will be remembered that when full-wave rectification was considered in connection with Fig. 37, the frequency of the rectifier pulses was double that of the supply mains.

In practice, the basic circuit shown in Fig. 42 is not very frequently employed owing to various difficulties that arise. For example, it is usually desirable to employ a filter of some kind to reduce the interference picked up from the mains wiring. In d.c. sets this interference is much greater than that usually experienced with a.c. mains. Furthermore, some kind of voltage stabilizing device is very often a big advantage and so is a means of by-passing stray high-frequency currents from the rectifier.

A circuit in which these modifications are employed is given in Fig. 43. In this circuit the mains are connected each to an h.f. choke. These chokes are shown at Ch_1 and Ch_2, and with C they form a filter to block the path of many types of interference coming along from the mains leads. Resistance R is for voltage dropping purposes and corresponds to the resistance R of Fig. 42. The ballast resistance R_2 is of a special type known as a barretter, i.e. its resistance increases when the current flow becomes greater than a given rating and decreases when the current flow diminishes. When the voltage from the mains increases the current flowing round the circuit would normally tend to increase. This increase in current makes the resistance offered by R_2 become greater and thereby compensates to a large extent for the increase in voltage provided by the mains. The rectifier valve is of the double-anode type and is indirectly heated.

The order in which the heaters of the receiver valves are connected is of some importance. It is generally desirable that the detector valve should be the last on the line and the order of the valves shown in Fig. 43 is as follows. Commencing at the rectifier heater, there are shown the heaters of the output valve V_5, intermediate-frequency amplifier V_4, h.f. amplifier V_3, frequency changer V_2, second detector V_1, and it is noted that V_1 is connected to the earth line.

For the provision of dial lights it is not usual to connect these across the heaters as is done in the case of a.c. valves. Instead, the lamps are joined in series with the heater circuit and across a resistance as shown in Fig. 43. If the lamps require less current

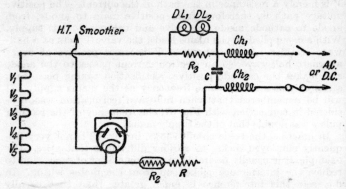

Fig. 43. An Alternative A.C./D.C. Mains Supply Circuit

than the valve heaters, R_3 is arranged to by-pass the excess current. In any event, a shunt resistance is desirable so that should a dial lamp become open circuited the heater circuit continuity is still complete.

Alternatively, the dial lamps could be connected together in parallel and then connected between the heater of V_1 and earth, or between the heaters of V_2 and V_1. In this instance, the current consumption of all the dial lamps should together equal that of one valve heater.

CHAPTER IX

COMPLETE MODERN RADIO RECEIVER CIRCUITS

This chapter is arranged to show how the circuits previously described are applied in actual practice. A selection of typical broadcast receiver circuits has been made that is representative of the modern trend in the broadcast receiver industry. These circuits will be found to embody the various stages already examined in the previous parts of this book. A study of these complete receiver diagrams, in conjunction with the notes on the individual stages, will give the reader a good idea of the circuit construction and operation of the great majority of modern broadcast receivers.

As most of the stages forming the receiver diagrams described here have already been discussed, a brief outline only of each receiver circuit is given. Where a new circuit is included, however, fuller details are given.

Three-valve Straight Battery Receiver. A circuit for this type of receiver is given in Fig. 44. Three pentodes are used, for in this way the maximum sensitivity is obtained. It will be readily appreciated that in a receiver employing only three battery valves the question of sensitivity is not an easy one to solve.

The aerial is coupled to the grid-cathode electrodes of the r.f. amplifier valve V_1 via a band-pass filter. This filter consists of two tuned circuits L_2 L_3 with variable condenser C_1, and L_4 L_5 with variable condenser C_2. These two circuits are coupled together by arranging the coils in the respective circuits to be fairly close so that electromagnetic interaction occurs between them. This is indicated by the broken line arrow passing through L_2 L_4 and L_3 L_5 respectively. The r.f. amplifier valve V_1 is of the tuned anode type, the tuned circuit being coils L_6 L_7, and variable condenser C_6.

Voltages from the tuned anode circuit are applied to the control grid of the grid detector pentode V_2 via the coupling condenser C_7 (about 0·0001 μF.), the grid leak being R_5 (2 megohms). The r.f. amplifier anode circuit is decoupled by R_4 (5 000 ohms) and C_4 (1 μF.). Reaction is employed with the detector, this being obtained by means of the coils L_8 L_9 coupled to the tuned anode coils of V_1. Control of reaction is by means of variable condenser C_5 (0·00035 μF.).

The pentode detector is operated with low anode and screen grid voltages. Screen grid feed from the h.t. positive line is

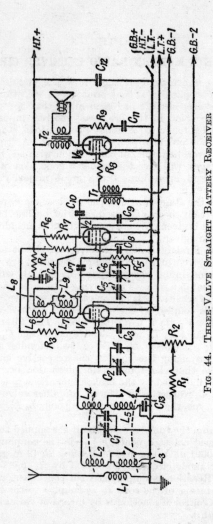

Fig. 44. Three-Valve Straight Battery Receiver

COMPLETE MODERN RADIO RECEIVER CIRCUITS

through R_6, which may be half a megohm in value and thereby drops the h.t. voltage down to perhaps only 12 volts. The anode load resistance may be about 100 000 ohms or more, reducing the anode voltage to about 35. Detector by-pass condenser C_9 should be used, but if it is made too high reaction will be insufficient. A value of 0·0002 µF. meets most requirements.

Parallel feed is used for the audio frequency transformer T_1, the coupling condenser being C_{10} (0·1 µF.). An h.f. stopper R_8 (50 000 to 200 000 ohms) is joined to the control grid of the output pentode V_3. This is connected in a conventional circuit with a fixed tone corrector consisting of R_9 (30 000 ohms) and C_{11} (0·01 µF.), the screen grid being joined directly to h.t. positive.

Across the h.t. leads is connected a reservoir condenser C_{12}. This may have a value of 2 µF. to 8 µF. and may be of the electrolytic type. Grid bias is provided by a battery, but automatic grid bias is practicable with battery receivers, one example of how this may be obtained being illustrated in Fig. 48. In the present case, the full bias voltage is applied to variable potentiometer R_2 (50 000 ohms), and the tapping on this potentiometer is joined to the control grid of the r.f. amplifier. By varying the position of the contact on the resistance element, a variable bias voltage is obtained. Valve V_1 is, of course, a variable mu valve, so by means of the adjustable potentiometer R_2, the amplification of the r.f. stage is varied. R_2 is, therefore, the volume control of the receiver. The bias circuit is decoupled by R_1 (2 megohms) and C_{13} (0·1 µF.).

Mains Three-valve Straight Receiver. It is usually desirable to have a band pass filter in the r.f. stage of a three-valve straight receiver in order to obtain the requisite amount of selectivity from the one r.f. valve. For maximum sensitivity a pentode grid detector and a pentode output valve are employed, although several three valve mains receivers use a triode in the output stage.

The diagram of a typical three-valve straight set is given in Fig. 45. The aerial is coupled by coils L_1 L_2 to a capacity coupled band pass filter, consisting of the two circuits L_3 L_4 C_2 C_1 and L_5 L_6 C_2 C_3. Condenser C_2 is the coupling condenser, this being common to the two circuits tuned by variable condensers C_1 and C_3 respectively. Resistance R_1 has a value of about 5 000 ohms.

Impedance or choke coupling is used between V_1 and V_2, the r.f. choke being L_7 connected to the anode of V_1. The voltage across the choke is applied to the tuned grid circuit of V_2 via C_6 (0·00005 µF.), the tuned grid circuit being L_8 L_9 and variable condenser C_8.

A potentiometer is used for providing screen grid voltage to V_1, comprising R_2 and R_3 R_4. The actual values given to these resistances will depend upon the required screen grid voltage. R_4 is used also as automatic grid bias resistance for V_1 by joining the cathode lead to the potentiometer tapping. By this means an effective control of amplification is provided. R_5 is to provide a minimum bias for V_1, so that even at the point of zero bias from R_4 a small negative bias is applied to the valve.

In order to provide high sensitivity, high values of grid condenser C_{10} (0·00025 μF.) and grid leak R_6 (2 megohms) for the detector may be used. Reaction is effected by means of coil L_{10} coupled to both coils L_8 L_9 in the tuned grid circuit, and is controlled by a differential condenser C_9. This type of condenser has one set of movable plates and two sets of fixed plates, one set of which is connected to chassis and the other to L_{10}. The rotatable plates are joined to anode. With a differential condenser as reaction control the effective capacitance across the detector valve does not alter appreciably with variation of control of reaction, because both sets of fixed plates or stators are joined between valve anode and chassis. As the rotor plates are moved from one set of fixed plates to the other the reaction varies, but the total anode-chassis capacitance between rotor and *both* stators does not.

The detector anode circuit is decoupled by R_8 (50 000 ohms) and C_{12} (4 μF.). This is advisable to prevent trouble due to motorboating. Anode load resistance is R_9 (250 000 ohms), the voltage from this being passed to the output valve via coupling condenser C_{14} (0·01 μF.). Grid resistance R_{10} (250 000 ohms) must not exceed the figure given by the valve V_3 manufacturers as being the maximum grid-cathode resistance for this valve. An anode stabilizing resistance R_{13} (100 ohms) is joined next to the anode of the output valve, which is connected in a conventional circuit with variable tone control C_{15} and R_{11}.

Full-wave rectification of the mains a.c. supply voltage is carried out in the mains supply equipment with rectifier V_4. Smoothing of the rectified voltage is by choke L_{11} and electrolytic condensers C_{18} (16 μF.) and C_{17} (16 μF.). In most receivers the loudspeaker field winding acts as smoothing choke, thereby avoiding the expense of a separate choke for this purpose. The valve heater winding is centre-tapped, the tapping being joined to chassis. At the primary of the mains transformer T_2 are three tappings which allow adjustments to be made to suit the different supply voltages in different localities.

Three-valve Mains Superheterodyne Receiver. The most usual design of a three-valve superheterodyne receiver consists of a frequency changer stage, intermediate frequency amplifier, and a double diode pentode for second detection, a.v.c., and as output valve. Owing to the high sensitivity of the modern double diode

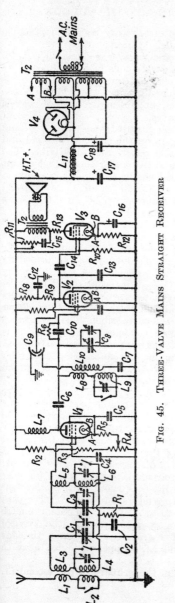

Fig. 45. Three-Valve Mains Straight Receiver

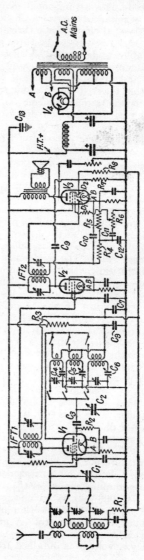

Fig. 46. Three-Valve Superheterodyne Receiver

pentode, this arrangement gives satisfactory results in practice, and is found in a number of models of broadcast receiver.

A typical circuit for a three-band receiver is shown in Fig. 46. The aerial is coupled to three separate coils which are tuned by variable condenser C_1 and switched in separately. Each coil is trimmed by a small condenser joined between the grid end of the coil and chassis. One aerial coupling is used on short waves, another for medium and long waves.

A triode hexode frequency changer is used, this valve providing a very satisfactory conversion gain on the short-wave band and helping the receiver to have a good overall sensitivity. The oscillator section (triode) of this valve uses a tuned grid circuit with variable condenser C_2. Grid condenser C_3 has a common value of 0·0001 μF., and grid leak R_2, 50 000 ohms, is returned to cathode. Note the position of condensers C_4, C_5, and C_6. These are in series with each tuning coil, and are for the purpose of ensuring that the tuned oscillator circuit is always correctly tuned with respect to the tuned input circuit of the hexode section of the valve to provide the proper intermediate frequency. We already know that the intermediate frequency (i.f.) is formed by combining voltages of different frequencies in the frequency changer valve. The condensers C_4 C_5 C_6 compress, as it were, the tuning band covered by variable condenser C_2, so that this difference frequency between the signal tuned circuits and the oscillator tuned circuits is always as correct as possible. The condensers referred to are known as padding condensers. This subject is rather subtle and a full explanation is not possible here. However, if the reader wishes to follow up the discussion, he should refer to *The Superheterodyne Receiver*, by A. T. Witts (Pitman), which covers the whole subject very clearly.

Two intermediate frequency transformers, i.f.t.$_1$ and i.f.t.$_2$, have both primary and secondary tuned by trimmer condensers. The usual i.f. is about 465 k.c./s, as this is found in practice to give the most satisfactory all-round results. Intermediate frequency amplifier valve V_2 is seen to be transformer coupled to the second detector diode D_1 forming part of V_3.

The detector circuit is of the type discussed in an earlier chapter, C_{10} (0·0001 μF.) being the reservoir condenser, R_4 (100 000 ohms) for i.f. filtering in conjunction with C_{12} (0·0001 μF.), R_5 (500 000 ohms) being the load resistance. Audio frequency voltages pass across coupling condenser C_{11} (0·01 μF.) to variable potentiometer R_6 (500 000 ohms) which acts as grid leak to the control grid of the pentode section of V_3, and at the same time as manual volume control. A grid r.f. stopper is joined next to this grid.

A.v.c. voltage is obtained from the primary of the second i.f.

transformer i.f.t.$_2$ via coupling condenser C_9 (0·00005 μF.), and the rectified a.v.c. voltage is produced across the load resistance R_7 of the a.v.c. diode D_2. This d.c. voltage, which is, of course, negative with respect to chassis, is applied via resistance R_8 (500 000 ohms) to the control grid circuits of V_1 and V_2, filter condensers being used at both these circuits.

The output stage of V_3 is the conventional one described in Chapter VI, and the mains supply equipment is similar to that used with Fig. 45. An additional condenser C_{13} across the smoothing condenser is employed in this instance. This has a capacitance of 0·1 μF. and is to provide a non-inductive by-pass across the rectifier circuit for r.f. currents.

Four-valve Mains Superheterodyne for A.C. or D.C.

In Fig. 47 is shown a diagram of the type of circuit to be expected in a four-valve superheterodyne receiver. Separate aerial coils are used for each waveband, and also separate input circuit coils tuned by variable condenser C_2. Connected across the aerial circuit is an acceptor circuit $L_1 C_1$ which is tunable by pre-set condenser C_1 to the intermediate frequency. This acceptor circuit offers a low impedance to currents at its resonant frequency, and accordingly any i.f. currents that are induced into the aerial circuit are by-passed across $L_1 C_1$ away from the receiver circuits proper. In this way, possible interference from stations transmitting on the intermediate frequency is prevented.

As frequency changer, a heptode is used with tuned oscillator grid circuit including variable condenser C_5. On medium and long wavelengths, two padding condensers C_6 C_7 and C_8 C_9 respectively are used. C_6 C_8 are fixed and C_7 C_9 are adjustable. By this arrangement the oscillator circuit is made less susceptible to drift of frequency due to alteration of the circuit constants. A much more constant capacity is provided by a suitable fixed condenser than by a pre-set condenser. Accordingly a proportion of the padding capacity (say one-half) is arranged to be supplied by a highly stable fixed condenser.

The oscillator and the input circuit medium and long waveband coils are short-circuited during reception on short waves by the switches shown across these coils. This prevents trouble due to absorption and other effects that are often experienced during short wave reception due to the proximity of resonant circuits. Condenser C_3 (0·00001 μF.) is sometimes used to neutralize an effect known as space charge coupling that may occur in a heptode or octode valve and becomes harmful on the short waves.

Variable selectivity is obtained by varying the coupling in the first i.f. transformer as indicated by the arrow through the coil symbol. As the coupling is reduced the selectivity is improved, but at the same time the reproduction suffers by a reduction in higher

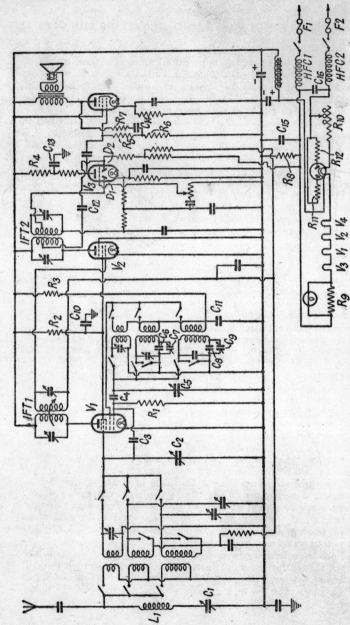

Fig. 47. Mains Four-Valve Superheterodyne

notes. The second transformer i.f.t.$_2$ coupling the i.f. valve V_2 to the diodes of V_3 is tapped at both primary and secondary. Such a tapping is used to reduce the damping produced by a diode and consequently improves the selectivity. In the present case, both the signal diode D_1 and the a.v.c. diode D_2 are tapped down the respective i.f.t.$_2$ windings for this purpose

Delayed a.v.c. is obtained by means of the bias applied from R_8, through which the total receiver h.t. current passes. As the a.v.c. diode D_2 has its load resistance joined to the end of the automatic grid bias resistance R_8 remote from the chassis, this negative bias is applied to the anode of D_2 and delays its action, usually by about 3 volts. At the same time the delay voltage is applied to V_1 and V_2 as a minimum bias via the a.v.c. circuit, and thereby renders unnecessary separate bias resistances in the cathode legs of these valves. Note that no a.v.c. is used on short waves.

Negative feed back is applied to the output valve V_4. Feed back voltage is passed through R_7 C_{14} to R_6 in the grid circuit. R_7 has to be chosen to be the correct ratio to R_6, and for a given value of R_7 the amount of food back is dependent upon the value of R_6. With most amplifiers a ratio of 10 : 1 for R_7 to R_6 will be found to give very good results, suitable values being 0·25 megohm for R_7 and 25 000 ohms for R_6. For C_{14} a value of 0·1 μF. may be used. Negative feed back reduces the distortion in the output valve to an extent dependent upon the proportion of output voltage feed back, and at the same time reduces the amplification correspondingly. In a superheterodyne receiver this diminished amplification can generally be made good by increasing the gain of earlier stages.

An a.c./d.c. mains equipment is employed, of the general type already described, and includes as mains h.f. filter HFC_1, HFC_2, and C_{16} (0·01 μF.), surge limiting resistances R_{11} R_{12} (50 ohms each) and dial lamp shunt R_9 (35 ohms).

Battery Five-valve Superheterodyne. The circuit shown in Fig. 48 is one commonly found in battery superheterodynes. Many of its essential features are the same as those found in mains receivers, the principal difference being the quiescent push-pull output stage which is favoured in battery superheterodynes as a means of economizing the consumption of h.t. current.

There are two connections for the aerial, A_1 for normal use and A_2 if interference is experienced. The connection of the aerial to A_2 inserts a small condenser C_1 (0·00002 μF.) in series with the aerial and thereby sharpens up selectivity. For short wave reception, the aerial circuit switch is closed and throws L_1 coupled to L_3 into circuit. During medium and long wavelength reception

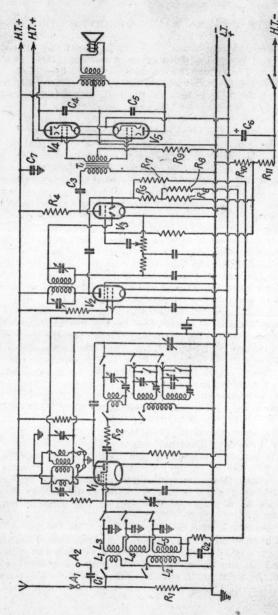

Fig. 48. Five-Valve Battery Superheterodyne

this switch is open and L_2 then becomes effective, this being coupled to L_4 and L_5. In addition to the inductive coupling across the coils, there is capacitative coupling by virtue of the condenser C_2 (0·004 μF.) being in the common leads to chassis of both the aerial circuit and the tuned input circuit. This arrangement thereby provides mixed inductive and capacitative coupling.

A triode pentode is used as frequency changer, with electron coupling. This is the modern version of the triode pentode described in Fig. 23. A resistance R_2 (100 ohms) is used to stabilize the operation of the oscillator section. In the first i.f. transformer i.f.t.$_1$ is a variable selectivity arrangement consisting of two small coils coupled to the primary and secondary respectively. The switching arrangement is such that the coils are short-circuited in alternative positions of the control switch mounted on the receiver panel. The effect of these coils is to decrease and increase the respective inductances of the primary and secondary windings and so to effect an alteration in the band width passed by the i.f. transformer, i.e. in the selectivity.

The i.f. stage is straightforward, and so is the signal diode detector. The a.v.c. diode is biased negatively for delayed operation by connecting the load resistance R_6 to the automatic grid bias resistances R_{10} R_{11}. Bias for the a.v.c. diode is obtained from R_{10}, which also biases the control grid of the triode section of V_3 owing to the connection of the grid leak R_3 being taken to it. The total d.c. voltage drop along R_5 R_6 is applied to the frequency changer V_1, while the drop across R_6 only is applied to the i.f. valve. This arrangement reduces the tendency to distortion that is present if the i.f. valve is heavily biased by the a.v.c. during reception of a strong signal.

Parallel feed is used for the audio frequency transformer T_1 coupling the triode of V_3 to the quiescent push-pull stage V_4 V_5. Often the two pentode valves of a quiescent push-pull amplifier are mounted in one envelope, as described in Chapter VII. In the present case, separate output tetrodes (beam valves) are employed, but the operation is similar to that given in the chapter referred to. The quiescent push-pull valves are biased from the biasing resistances R_{10} R_{11}, a stabilizing resistance R_9 (150 000 ohms) being in the common grid return lead.

Six-valve Mains Superheterodyne. The use of an r.f. amplifier in a superheterodyne receiver is a great advantage. Quite apart from the increased sensitivity, an r.f. stage is justified on the grounds of reduced background noise, improved rejection of "image" or second channel signals, and adjacent channel selectivity.

In Fig. 49 the r.f. stage is a straightforward transformer coupled amplifier, a resistance R_1 being joined across the long wave coil

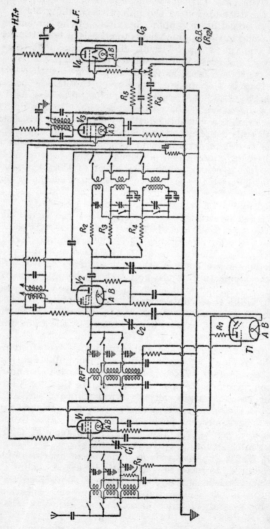

Fig. 49. Circuit of Six-Valve Superheterodyne Receiver, up to Second Detector and A.F. Amplifier

COMPLETE MODERN RADIO RECEIVER CIRCUITS

to damp this circuit and avoid sideband cutting due to excessive selectivity.

For levelling up the oscillator voltage generated by the triode section of the frequency changer V_2 there are connected three resistances, R_2 R_3 R_4, one for each waveband. Typical values are: short waveband R_2, 350 ohms; medium waveband R_3, 1 000 ohms; long waveband R_4, 2 000 ohms. The use of these resistances provides a more uniform value of oscillator voltage and consequently a more equal conversion gain and overall sensitivity at the different points of the tuning scale.

The i.f. transformers are trimmed by adjustable iron cores, indicated by the arrow through each coil symbol. The i.f.t. condensers are, therefore, of fixed capacitance.

The diodes of V_4 are joined together (strapped diodes) and jointly act as combined detector and a.v.c. valve. From the load resistance R_5 (500 000 ohms) the a.v.c. voltage is taken via R_6 (50 000 ohms), and the audio signal is passed over the coupling condenser to the control grid of the triode of V_4. A condenser C_3 (0·0002 μF.) is used to by-pass any i.f. currents that might be present in the output of this valve.

As output stage a push-pull amplifier is used as shown in Fig. 50. This is coupled by an auto transformer to the a.f. amplifier V_4. Resistances R_7 R_8 (50 000 ohms) prevent parasitic oscillation. An extension loudspeaker may be joined across the output transformer secondary winding, and when this is used the switch S may be opened to silence the main loudspeaker. A coil L_1 in series with the speech coil is coupled to the loudspeaker field winding L_2 acting as smoothing choke and has induced into it a hum voltage in opposite phase to that already there. Coil L_1 is known as the hum bucking coil, and is so designed that the net hum frequency voltage is zero.

The mains supply equipment is straightforward, except in that there are two heater windings for the receiver valves. Terminals AB supply heater current to all stages except the output, which is supplied by CD. It will be noted that the push-pull stage has directly heated valves. To provide bias, a resistance R_{11} is joined from the centre point of R_{10} to chassis. This common bias resistor need not be by-passed by a condenser for a push-pull amplifier. Instead of centre tapping the heater windings, resistances R_9 and R_{10} are joined one across each winding and the centre point of these resistances are taken to chassis. These resistances, about 30 ohms in value, are often referred to as hum dingers.

A tuning indicator TI is commonly used in modern receivers. This consists of a triode amplifier, to which the a.v.c. voltage is applied, and a cathode ray indicator. The control electrode of the indicator section is joined inside the envelope to the anode

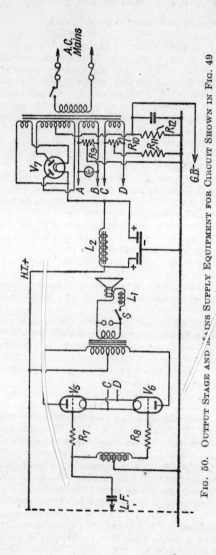

Fig. 50. Output Stage and Mains Supply Equipment for Circuit Shown in Fig. 49

of the amplifier triode, which is connected to the h.t. + line through R_7 (1 megohm). The anode of the cathode ray tube acts as target on which electrons strike and set up fluorescence over a controlled area. As the a.v.c. bias increases, the area of shadow on the fluorescent screen is reduced. The correct tuning point is thus that corresponding to minimum shadow angle.

MODERN RADIO COMMUNICATION

By J. H. REYNER, B.Sc. (Hons.), A.C.G.I., D.I.C., A.M.I.E.E., M.Inst.R.E. With a Foreword by PROFESSOR G. W. O. HOWE, Wh.Sch., D.Sc., M.I.E.E.

In two vols. Each crown 8vo, cloth gilt, illustrated.

Vol. I, 334 pp. 7s. 6d. net.

This book has been thoroughly revised, and now includes all recent developments, with full descriptions of the latest apparatus and instruments. At the same time it provides a solid groundwork of essential fundamental principles. Although primarily intended for the Preliminary and Intermediate Grades of the City and Guilds Examinations and the P.M.G. Certificate, it will be found helpful by practical engineers and ranks with many large and costly treatises.

TECHNICAL JOURNAL: "*The book is well written and accurate, and the elementary technological student will find it of great assistance and very readable.*"

Vol. II, 255 pp. 7s. 6d. net.

Specially written to assist students in preparing for the Final Examination in Radio Communication of the City and Guilds of London Institute. With its companion volume it covers every aspect of the subject, and provides a complete groundwork in the theory and practice of Modern Radio Engineering.

PITMAN

INDEX

ACCEPTOR, circuit for i.f., 93
Amplification, 1
——, h.f., 1
Anode, detector, 15
—— dissipation, 55
Audible limit, 23
Automatic grid bias, 82
A.V.C., delayed, 35, 95, 97
—— filter, 8, 33
——, quiet, 37
—— theory, 32

BALLAST resistance, 85
Band pass filter—
 advantage of, 4
 capacitively coupled, 10, 89
 inductively coupled, 7
 mutual inductance coupling, 4

CATHODE injection, 49
Centre tapping for heater winding, 78
Choke coupled amplifier, see IMPEDANCE COUPLED AMPLIFIER
—— for a.c./d.c. mains supply, 85
—— for output valve, 56
——, h.f., 5
——, smoothing, 81
——, use in detector circuit, 14
——, —— in diode circuit, 36
Class A push-pull, 62
—— B push-pull, 64
Condensers—
 a.v.c. filter, 33
 coupling, band pass filter, 10
 ——, h.f., 5
 ——, l.f., 24, 26
 decoupling detector, 17
 detector by-pass, 14
 differential, 90
 diode load circuit, 20
 grid detector, 15
 screen grid, 3, 18
 smoothing, 81, 82
 trimmer, 4, 18

DECOUPLING, 17
——, grid circuit, 42
Degeneration, 74, 83

Detector anode, 15
——, diode, 19
——, grid, 13
——, pentode, 17
—— theory, 11
Dial lamps for a.c./d.c. sets, 86
—— —— for a.c. sets, 79
Diode detector, 19
—— —— for a.v.c., 32, 35
—— ——, push-pull, 20
—— ——, strapped anodes, 33
Distortion, Class B, 66
—— in l.f. amplifiers, 70
—— with automatic grid bias, 84
Double-diode-pentode circuits, 35, 38, 39
Double-diode-triode circuit, 33

FEED-BACK in h.f. amplifiers, 2
Filter, h.f., 6
Five-valve superheterodyne, 95
Four-valve superheterodyne, 93
Frequency changers, heptode, 44
—— ——, theory, 43
—— ——, triode-hexode, 49
—— ——, triode-pentode, 48
Fuses, use of, 78

GRID bias resistance, minimum, 5
—— —— ——, sources of, 82
—— —— ——, theory of working, 82
Grid detector, 13
—— —— values of condenser and leak, 15

HARMONICS in output stage, 54
—— suppressor, 49
Heater circuit, a.c., 77
—— ——, a.c./d.c., 84
Heptode circuit, 44
Heterodyne filter, 61
Hexode, definition of, 49
H.F. stoppers, diode choke, 36
—— ——, diode circuit, 33
—— ——, l.f. circuit, 58
—— ——, mains supply circuit, 77, 86
H.T. battery for Class B, 67

H.T. battery, voltage and power output, 55
Hum in a.c. receivers, 80

IMPEDANCE coupled amplifier—
h.f., 3, 89
l.f., 30
Instability, heptode circuit, 46, 47
——, h.f. amplifiers, 2
——, ——, preventing, 5, 8, 10
——, quiescent push-pull, 69

LEAKAGE inductance, 27
Low frequency amplifier, operating conditions, 22

MAINS aerial, 79
—— hum, 80
Matching output valve and loudspeaker, 57
Metal rectifier circuit, 79
—— v. valve, 75

NEGATIVE feedback, A.F. 95
Noise suppression, 37

OCTODE circuit, 47
Oscillator, condenser and leak resistance, 45
—— stabilizing resistances, 97, 99
Output stage, 54

PARALLEL feed, l.f., 28, 89
—— —— oscillator, 52
—— —— output valve, 56
Paraphase push-pull, 70, 72
Pentode detector, 17, 87
—— output valve, 57
Phase changer, 71, 73
Power output, 55
—— supply circuits, 75
Push-pull amplification, 62
——, Class A, 62, 99
——, —— B, 64
——, detector, 20
——, quiescent, 68

QUIESCENT push-pull amplifier, 68

REACTION feed current, 12
Reaction process, 14
—— resistance, 10
Rectifier valve circuit, 75
—— —— v. metal, 75
Rejector, h.f., 6
Reservoir condenser, 81
—— ——, battery sets, 89

Resistance, amplifier anode, 25
——, a.v.c. filter, 20, 33
——, decoupling, 17
——, grid circuit stabilizer, 10
——, —— leak for detector, 15
—— pentode-anode, detector, 18
——, reaction circuit, 10
——, screen grid decoupling, 8
Resistance-capacitance amplifier, 23

SCREEN grid condenser, 3, 18
—— —— potentiometer, 5, 89
—— —— voltage in output valve, 59
Smoothing choke, 81
—— circuit, 80
—— condenser, 81, 82
Superheterodyne receiver, 44, 92

THREE-VALVE straight receiver:
battery, 87
—— —— ——, mains, 89
—— —— —— superheterodyne receiver, 90
Tone control, 34
—— ——, Class B, 67
—— —— in input circuit, 60
—— —— in output circuit, 61
—— —— pentode, 58
—— ——, quiescent push-pull, 69
Transformer coupled amplifier—
h.f., 6, 7
i.f., 6
l.f., 26
parallel feed, 28
Triode-hexode, 49, 92
Triode-pentode frequency changer, 48, 97
—— i.f. and l.f. amplifier, 40
—— anode circuit, 8, 10
—— —— oscillator, 48, 52
Tuned grid circuit, 4
—— —— oscillator, 45, 50
Tuning indicator, 99

UNIVERSAL valves, power supply for, 84

VARIABLE, selectivity, 93
Voltage doubler circuit, 80
Volume control, automatic, see a.v.c.
—— ——, manual, 34, 89